Enhanced Geothermal Systems (EGS)

Peter Meisen, Past President, Global Energy Network Institute, asked in 1997, "What if there was an existing, viable technology, that when developed to its highest potential could increase everyone's standard of living, cut fossil fuel demand and the resultant pollution?" After 23 years of sustained effort by the global scientific community, this is becoming a reality.

The technology to extract heat from granite has been revolutionized in the last few years. The classical method of creating fracture networks by hydrofracturing is being replaced by a closed-loop method where fluids are not in contact with the hot granite. Supercritical CO_2 is replacing water as a circulating fluid. Certainly, the future energy road is going to be led by highly radiogenic granites. While hydrothermal sources are site-specific and have their limitations, EGS can be initiated anywhere on earth. EGS is removing all such obstacles and, in the future, will provide uninterrupted electricity for all. Energy-deficient countries can have surplus electricity; water-stressed countries can have a perennial freshwater supply; and countries can become food-secure and rise above poverty levels. Countries need not depend on energy imports and can independently evolve into carbon neutral or low carbon societies.

The contributions made by experts will help researchers and investors to close the energy demand and supply gap in the very near future by tapping the unlimited energy of the Earth.

Opportunities available for investors in Turkey are well documented with field, geophysical, and geochemical data and information on the energy generating capacity of the granite intrusive spread over a cumulative area of 6,910 km^2 in western Anatolia.

With the signing of the Global Geothermal Alliance (GGA) by several countries during the December 2015 CoP 21 (Conference of Parties) summit in Paris, countries are obliged to reduce CO_2 emissions by increasing the footprint of renewable energy in the primary source mix. Information provided in this book will lead the way to establishing a clean energy future for millions of people for sustainable development and help to mitigate crises arising due to food, water, and energy shortage issues. Academic and research institutes will benefit to a large extent from the expertise of the top contributors in this book. This information provided in this book will help to lay the foundation for super-hot EGS research in future.

Enhanced Geothermal Systems (EGS)

The Future Energy-Road Ahead

Edited by

Dornadula Chandrasekharam
and Alper Baba

CRC Press
Taylor & Francis Group
Boca Raton London New York Leiden

CRC Press is an imprint of the
Taylor & Francis Group, an **informa** business
A BALKEMA BOOK

Designed cover image: Dornadula Chandrasekharam

First published 2024
by CRC Press/Balkema
4 Park Square, Milton Park, Abingdon, Oxon, OX14 4RN
e-mail: enquiries@taylorandfrancis.com
www.routledge.com – www.taylorandfrancis.com

CRC Press/Balkema is an imprint of the Taylor & Francis Group, an informa business

ISBN: 978-1-032-22185-4 (hbk)
ISBN: 978-1-032-22188-5 (pbk)
ISBN: 978-1-003-27147-5 (ebk)

DOI: 10.1201/9781003271475

Typeset in Times New Roman
by codeMantra

Contents

Editor biographies

Dornadula Chandrasekharam (Chandra: 1948, India), TUBITAK Fellow, Izmir Institute of Technology (IYTE), Turkey, obtained his MSc in Applied Geology (1972) and PhD (1980) from IITB. He has been working in the fields of geothermal energy resources, volcanology and groundwater pollution for the past 45 years. Before joining IITB, he worked as a Senior Scientist at the Centre for Water Resources Development and Management and the Centre for Earth Science Studies in Kerala, India, for 7 years. he held several important positions during his academic and research career. Before joining IYTE, He served as a Chair Professor, Indian Institute of Technology Bombay (1987–2015) and Visiting Professor, Indian Institute of Technology Hyderabad (2016–2021). He was a TWAS Visiting Professor to Sanaa University, Yemen Republic (1996–2001); Senior Associate of Abdus Salam International Centre for Theoretical Physics, Trieste, Italy (2002–2007); Adjunct Professor, China University of Geosciences, Wuhan (2011–2012), Visiting Professor to King Saud University of Saudi Arabia (2011–2017) and Adjunct Professor, University of Southern Queensland, Australia (2016–2019). He received the International Centre for Theoretical Physics (ICTP, Trieste, Italy) Fellowship to conduct research at the Italian National Science Academy (CNR) in 1997. Prof. Chandra extensively conducted research in low-enthalpy geothermal systems in India and is currently the Founder Chairman of M/s GeoSyndicate Power Private Ltd., the only geothermal company in India. He was an elected board member of the International Geothermal Association (2010–2016) and has widely represented his country at several international geothermal conferences. He conducted short courses on low-enthalpy geothermal resources in Argentina, Costa Rica, Poland, China and Turkey. He has supervised 25 PhD students, published 210 papers in peer-reviewed journals of international repute and published 7 books in the fields of groundwater pollution and geothermal energy resources. His books on geothermal energy resources (i) *Geothermal Energy Resources for Developing Countries* (Balkema Pub., 2002), (ii) *Low Enthalpy Geothermal Resources for Power Generation* (Taylor and Francis, 2008), (iii) *Geothermal Systems and Energy: Turkey and Greece* (CRC Press, 2014) and (iv) *Red Sea Geothermal Provinces* (CRC Press, 2018) are widely read. Prof Chandra served on the Board of Director of (i) Oil and Natural Gas Corporation, (ii) Western Coal Fields Ltd., (iii) Indian Rare Earths Ltd. and (iv) Mangalore Refineries and Petrochemicals.

He has been appointed as the Chairperson of the Geothermal Energy Resources and Management committee constituted by the Department of Sciences and Technology, Government of India. Currently, he is with the International Water Resources Department, IYTE, Turkey.

Alper Baba (Turkey, 1970) holds a degree in geology and a doctorate in the field of hydrogeology from Dokuz Eylul University, Izmir. He has 25 years' experience in hydrogeological and environmental geology problems in different parts of the world. Since 2010, he has been a professor at the Izmir Institute of Technology as the director of the International Water Resources Department. He is teaching and conducting research in the fields of environmental geology, groundwater contamination, geothermal energy and hydrogeology. He has coordinated a variety of national and international R&D projects in cooperation with research institutes and companies, among them NATO, World Bank and EBRD-funded projects. Dr. Baba was awarded the "Successful Young Scientists Award" of the Turkish Academy of Science and the "Gold Medal Award" of the Turkish Geological Engineering Association Dr. Baba is the author of several peer-reviewed scientific publications and contributions to international conferences and is the editor of the books "Groundwater and Ecosystems," "Climate Change and its Effects on Water Resources, Issues of National and Global Security" (NATO Science Series, Springer) and "Geothermal Systems and Energy: Turkey and Greece" (CRC Press). Currently, he is the Vice Rector of the Izmir Institute of Technology, Izmir.

Contributors

Corinna Abesser
BGS
Nottingham, United Kingdom

Hiro Asanuma
FREA
Fukushima, Japan

Tolga Ayzit
International Water Resources Department
Izmir Institute of Technology
Izmir, Turkey

Alper Baba
International Water Resources Department
Izmir Institute of Technology
Izmir, Turkey

Roy Baria
Mil-Tech UK Ltd
Chertsey, United Kingdom

Clément Baujard
Geosciences Department
ES-Géothermie
Strasbourg, France

Joerg Baumgaertner
BESTEC GmbH
Berlin, Germany

Graeme Beadsmore
Hot Dry Rocks Pty
Carlton, VIC, Australia

Dornadula Chandrasekharam
International Water Resources Department
Izmir Institute of Technology
Izmir, Turkey

A. Chaudhuri
Department of Applied Mechanics
Indian Institute of Technology Madras
Chennai, India

Eléonore Dalmais
Geosciences Department
ES-Géothermie
Strasbourg, France

Silvia De Simone
Géosciences Rennes
University of Rennes, CNRS
Rennes, France

Albert Genter
Geosciences Department
ES-Géothermie
Strasbourg, France

Carole Glaas
Geosciences Department
ES-Géothermie
Strasbourg, France

Hylke Glass
Camborne School of Mines
Penryn, United Kingdom

F.J. Huang
Faculty of Infrastructure Engineering
Dalian University of Technology
Dalian, China

Shahin Jamali
Advanced Drilling Group
Fraunhofer IEG
Bochum, Germany

K. Ji
Faculty of Infrastructure Engineering
Dalian University of Technology
Dalian, China

Andrew Jupe
Altcom Ltd
Penzance, United Kingdom

Iman Rahimzadeh Kivi
Institute of Environmental Assessment and Water Research
Spanish National Research Council
Barcelona, Spain

Igor Kocis
GA Drilling
Bratislava, Slovakia

W.G.P. Kumari
School of Civil, Mining, Environmental and Architectural Engineering
Faculty of Engineering and Information Sciences
University of Wollongong
Wollongong, NSW, Australia

H. Li
Faculty of Infrastructure Engineering
Dalian University of Technology
Dalian, China

Y. Liu
Faculty of Infrastructure Engineering
Dalian University of Technology
Dalian, China

Vincent Maurer
Geosciences Department
ES-Géothermie
Strasbourg, France

Ki-Bok Min
Department of Energy Resources Engineering
Seoul National University
Seoul, South Korea

Francesco Parisio
Global Change Research Group (GCRG)
Spanish National Research Council (IMEDEA, CSIC-UIB)
Esporles, Spain
and
Institute of Environmental Assessment and Water Research
Spanish National Research Council (IDAEA-CSIC)
Barcelona, Spain

Berker Polat
Advanced Drilling Group
Fraunhofer IEG
Bochum, Germany

P.G. Ranjith
Department of Civil Engineering
Monash University
Melbourne, VIC, Australia

Guillaume Ravier
Exploitation, Engineering and Maintenance Department
ES-Géothermie
Strasbourg, France

Ann Robertson-Tait
GeothermEx
Richmond, California

Jonny Rutqvist
Lawrence Berkeley National Laboratory
Berkeley, California

M. Singh
Group of Geothermal Science and Technology
Institute of Applied Geosciences
Technical University of Darmstadt
Darmstadt, Germany

Sri Kalyan Tangirala
Department of Applied Geophysics
Indian Institute of Technology (Indian School of Mines)
Dhanbad, India
now at
Global Change Research Group (GCRG)
Spanish National Research Council (IMEDEA, CSIC-UIB)
Esporles, Spain

Y. Tao
Faculty of Infrastructure Engineering
Dalian University of Technology
Dalian, China

Dimitra Teza
BESTEC GmbH
Berlin, Germany

Victor Vilarrasa
Global Change Research Group (GCRG)
Spanish National Research Council (IMEDEA, CSIC-UIB)
Esporles, Spain

Volker Wittig
Advanced Drilling Group
Fraunhofer IEG
Bochum, Germany

Foreword

In rural areas, greater than 2 billion people in the world today live below the poverty line without adequate energy, water and food. These are the people who are severely affected by climate-related issues and face extreme weather conditions. These are the people who live in regions that are experiencing water stress. It has been calculated that 7% of total world electricity generated could protect them and provide basic human needs. In the current century, with advanced technological and management skills, these people's conditions are not better than two decades ago. What we need to overcome this agony is a viable technology that, when developed to its maximum potential, could provide energy, food and water security to everyone, increase their living standards, especially the rural population, reduce pollution and offset the major use of fossil fuels. We have reached the stage where the shortage of energy has become a myth. Clean and renewable energy is abundant now but only marginally utilized. Today, technology has matured to tap this green energy and control global climate-related disasters. This energy is locked up in the crust, in the radiogenic granites that are abundant at any place on earth. Then it was christened as the "hot dry rock" geothermal resource and is now termed as Enhanced Geothermal Systems (also termed as Engineered Geothermal Systems or Advanced Geothermal Systems). Advanced drilling technology has revolutionized the energy sector. This technology of harnessing heat from the granite has come as a shot in the arm to countries across the world. Just a km^3 of such granite has the potential to supply 79 million units of electricity and a copious amount of heat that can be utilized in a variety of ways humans require it to heat and cool homes, grow food in green houses and generate freshwater from the sea. This is the main focus of sustainable development that countries need. This energy is not site-specific. It is there in rural and urban areas! When the full potential of the granites is developed, countries can be energy-food-water independent.

This book, with contributions from scientists in the EGS field of research who soiled their hands during the development of this technology, is brought out at a time when countries are looking for viable, carbon-free energy that can support baseload power supply and help millions to rise above the poverty line.

I compliment the editors for bringing out this important book at a time when countries have become signatories to the CoP 2015 Paris Agreement. I am sure this technology will lead the countries towards achieving the net zero emissions (NEZ) target by the end of 2040.

Marit Brommer
Executive Director
International Geothermal Association

Foreword

Under the present context of global warming due to excess emissions of GHG, countries should give high priority to developing renewable energy resources, even if the process is challenging. Renewable energy technologies will help the poor to achieve the benefits of development without having to face the same environmental costs the developed world has experienced. Energy and water are becoming the most wanted commodities in the growing world today.

Every country is trying hard to be energy, water and food independent. It is fortunate that Turkey is endowed with huge geothermal energy potential and has already made a mark on the world geothermal map. While the use of hydrothermal sources is on the rising path, contributing immensely to the country's sustainable developmental programmes, the country's highly radiogenic granite, spread over an area of 6,910 km^2, is waiting to be explored and exploited. Enhanced Geothermal Systems, a technology that harnesses the heat from such granites, is the future energy for mankind for solving issues related to climate change due to excess emissions of CO_2. With energy security comes food and water security. Today, due to emerging technological developments in drilling, harnessing the heat from the granites is becoming a viable task. The classical method of hydrofracturing granites to mine the heat is being replaced by a closed-loop heat extraction method that eliminates the risk of minor seismicity. Supercritical carbon dioxide is replacing water as heat extraction fluid. Thus, CO_2 sequestration has become a part of the energy harnessing method in EGS.

The technology to generate electricity, produce freshwater, support greenhouse cultivation and preserve food products through dehydration using geothermal energy is well developed. Turkey stands third in the world in such technologies. The greatest advantage of geothermal energy is that the energy supply is not influenced by external factors like rain, wind and other extreme events. Unlike other renewables, it is online 90% of the time with a low land and CO_2 footprint.

This book is being published at a time when Turkey has become a signatory to the Paris 2015 Agreement to reduce CO_2 emissions by increasing the quantum of renewable energy in the primary energy source mix and reducing gas imports.

The authors in this book brought out excellent mitigation strategies to circumvent CO_2 emission-related issues by highlighting the technological developments involved in EGS. I congratulate the editors for having brought out this excellent book at a time when it is needed. This book, I am sure, will be very useful to graduate students, researchers and institutes involved in renewable energy research. The book contains a wealth of information for energy policymakers, administrators and planners in all countries to plan mitigation strategies to fight the problems related to global warming and climate change.

Yusuf Baran
Rector
Izmir Institute of Technology, Izmir

Preface

The world is sailing through an energy crisis due to several geopolitical factors while this book is getting into the press. Countries are preparing for the next CoP 27, being hosted by Egypt, with the hope of bringing about a major transition in energy usage by emerging economies around the world. While hydrothermal sources are making a greater contribution to support the energy transition, Enhanced Geothermal Systems or Engineered Geothermal Systems (some term it Advanced Geothermal Systems or Petrothermal Systems) are emerging, as the title of this book rightly suggests, as the strong future primary energy source. The reason for EGS to be the future sustainable energy source is that hydrothermal systems depend on natural water circulation and are hence site-specific and thus confined to subduction zones and major continental rift systems, while EGS has no such site restrictions and depends entirely on the heat availability at any site. Furthermore, the land footprint of EGS is very small compared to all other renewable energy sources.

Advancements made in drilling technologies are making the levelized cost of power from EGS more affordable for the countries. Directional drilling, loop technology, well-in-well technology and down-the-hole heat exchangers are being explored by several companies and research laboratories across the world to minimize the cost of drilling and reduce the induced seismicity that is closely associated with conventional EGS technology. The chapters in the book, written by experts in the field of EGS, demonstrate that the countries are at the threshold of major clean energy source technology breakthroughs. While Earth is a huge source of energy, not even 1% of its potential is used to meet the energy demand. Geothermal energy being a contributor to the primary energy source and saving CO_2, the reservoirs can also be used for CO_2 sequestration, as explained using numerical models (Chapter 1). The first EGS power plant that is successfully put into operation in Soultz-sous-Forêts (SsF) is able to save about 36 g CO_2-eq/kWh. The SsF power plant demonstrated that geothermal energy could produce renewable electricity, heat at high temperatures for industry and tomorrow's neutral carbon lithium for batteries. (Chapter 2). The main source for EGS is granites with high radiogenic element content. Such granites are available anywhere on the earth, and hence, EGS is termed as "energy below your feet." Western Anatolia is a warehouse of such granites spread over an area of 6,910 km^2, with an estimated capacity to generate 546×10^9 kWh of power. Besides power, this energy can be utilized to generate freshwater through the desalination process at a cost of <1.5 euros/m^3 of freshwater generated, thus providing energy and water security to the country (Chapter 3). Like any other new technology, EGS too has several bottlenecks and methods to circumvent them. A new scheme known as EGS-E is proposed for the exploration of large-scale heat from the hot rock source (Chapter 4).

Since EGS involves nonporous reservoir rocks, reservoir stimulation, which includes hydraulic, chemical and thermal stimulation technologies, is very essential. The hydraulic stimulation technology extensively used for fracturing oil and gas wells has been widely adopted in EGSs. Unlike oil and gas reservoirs, EGSs are associated with high-temperature and high-pressure regimes, and hence, it is essential to consider the thermal stresses associated with hydraulic fracturing. This can be addressed using supercritical CO_2 as a circulating gas, which is giving promising results. Field trials are being attempted currently (Chapter 5). The most challenging issue associated with EGS reservoir development is induced seismicity. A clear understanding of such issues is essential. Numerical simulation models play an important role in understanding the interaction between fluid flow, stress variations, heat transport and geochemical reactions. Extensive research is underway to address these issues currently, and the results are being implemented in several ongoing EGS projects across the world (Chapter 6). Injection protocols in EGS reservoirs play an important role in understanding fracture propagation and the hydraulic connection between wells. These include dilatancy, angle of dilatancy, influence of stress regime and fluid injection rate. These protocols, in particular cyclic injection protocols, in conjunction with thermal reservoir simulation and flow-related stress regimes, play an important role in the EGS development (Chapter 7). A common problem with the EGS is the impediment of connectivity between the injection and production borewells due to mud clogging or other similar processes. These fractures thus need "repair" to re-establish the fracture connectivity, normalize the fluid flow and increase the productivity. This can be accomplished using radial jet drilling, which is being used in hydrocarbon wells, and this technology is being developed to support EGS (Chapter 8). Radiogenic granites occur over a wide geographic region of the world, and hence, a large number of scientists, engineers and technologists in different institutes and research laboratories are involved in research related to EGS. It is important to bring them together on one platform to exchange field and laboratory experience. This will help the geothermal group to troubleshoot minor issues arising during the exploration and exploitation of the heat from deep underground. Chapter 9 accomplishes this task, and this book provides a sound background on the various laboratories involved in EGS research to smooth the energy road for sustainable development and mitigate issues that arise due to the excess release of GHG into the atmosphere.

Dornadula Chandrasekharam and Alper Baba
Izmir Institute of Technology, Turkey

Acknowledgements

The editors sincerely thank Abesser, C, Ann, R.T., Asanuma, H., Ayzit, T., Bariya, R., Baumgaertner, J., Beadsmore, G., Baujard, C., Chaudhuri, A., De Simone. S., Dalmais, E., Genter, A., Glaas, C., Glass, H., Huang, F.J., Jamali, S., Jupe, A., Ji, K., Kivi, I.R., Kocis, I., Kumari W.G.P., Li, H., Maurer, V., Min, K. B., Polat, B., Parisio, F., Ravier, G., Ranjith. P.G., Rutqvist, J., Singh, M., Tangirala, S.K., Tao, Y., Teza, D., Vilarrasa, V. and Wittig, V, for their patience and support during the preparation of this book. We appreciate the cooperation of some of the authors, who, in spite of their ill health due to the COVID pandemic, were able to meet the deadline.

The editors place on record the invaluable cooperation and support extended by the reviewers, who readily accepted to review the manuscript, which has improved its quality.

The editors thank Prof Dr Yusuf Baran, Rector, Izmir Institute of Technology, Izmir for his constant support and encouragement during the course of this work.

This book is part of the Project Entitled "ENHANCED GEOTHERMA SYSTEMS: EXTRCATING ENERGY FROM GRANITES FOR ENERGY AND FOOD SECURITY OF TURKEY" (Project No: 120C079) under TUBITAK 2236 Co-Funded Brain Circulation Scheme 2. The editors place on record the support of TUBITAK during the course of this work.

SPECIAL APPRECIATION TO JANJAAP BLOM

Our association with Janjaap Blom started in 2002. Over the last 20 years, we have never had to look for an alternate publishing house to publish our eight books. This speaks volumes of the cooperation, understanding and faith we got from CRC Press/Balkema under the able leadership of Janjaap Blom. Our ninth book coincides with the forthcoming retirement of Janjaap Blom. It is sad that our colleague is retiring from his position...but one has to accept that it is part of life. We thought, on this occasion, to place on record the cooperation, guidance and support we enjoyed from Janjaap Blom. We cherish the memories we have enjoyed over the last 22 years. We wish him a fruitful future, enjoying life with his grandchildren. We miss you, Janjaap!

Dornadula Chandrasekharam and Alper Baba
Izmir Institute of Technology, Turkey

Chapter 1

Evaluation of low- to moderate-enthalpy shallow sedimentary reservoirs for CCS-CPG systems

M. Singh
Technical University of Darmstadt

A. Chaudhuri
Indian Institute of Technology Madras

CONTENTS

1.1 INTRODUCTION

CO_2 from fossil-fired electricity generation plants (Mohan et al., 2015; Davidson et al., 2017) and various other heavy industries such as steel, cement, petrochemicals, etc. can be utilized as the working fluid for geothermal energy extraction to offset the cost of carbon capture and storage (CCS). Geothermal energy is available in hydrothermal sedimentary (HS) reservoirs and hot dry rock (HDR) systems. Even though the global potential of HDR systems is significantly larger than that of HS reservoirs, energy extraction from HDR systems is very challenging because they are usually very deep and very less permeable (Garapati et al., 2017). HS reservoirs are present at shallow depths, and these also possess higher porosity and permeability. Therefore, these can be suitable for geothermal energy extraction through convective heat extraction mechanisms and CO_2 sequestration (Thorsteinsson et al., 2008; Randolph and Saar, 2011; Garapati et al., 2017; Pan et al., 2019). Presently, such reservoirs are getting attention due to their unique power production, resource optimization (Gosnold, 2015), and proximity to fossil fuel-based power plants (Davidson et al., 2017). Some notable existing geothermal reservoirs are in the Raft River geothermal valley (reservoir temperature is below 145°C at depths of 1,524–1,829 m), Cove Fort geothermal area (mean reservoir temperature is 150°C with depth varying between 180 and 1,195 m) in the USA, and Paralana geothermal system in Australia (Brugger et al.,

DOI: 10.1201/9781003271475-1

2005) (reservoir temperature is below 95°C at depths of 14–2.4 km). There are more than 1,000 geothermal sites in the USA with a reservoir temperature up to 90°C that are categorized as low-enthalpy reservoirs as listed by the USGS (Reed, 1983). Indian geothermal reservoirs (e.g., Sonata, Combay, Godavary, and Mahanadi basins) are of low enthalpy with temperatures less than 100°C (Yadav and Sircar, 2020) and are present in the close vicinity of the existing coal- based thermal power plants.

For CO_2 plume geothermal (CPG) systems in HS reservoirs, the displacement of brine by injected liquid or supercritical CO_2 is a two-phase flow process (Randolph and Saar, 2010). Randolph and Saar (2011) first simulated two-phase fluid flow for a horizontal domain with the five-spot well pattern, and they showed the advantage of the higher mobility of CO_2 than water in extracting energy at a higher rate. The effect of the density difference between brine and CO_2 was not included in their study for a horizontal domain. Pan et al. (2018) performed 3D numerical simulations of CPG systems. The effect of buoyancy was clear from their results of an evolving CO_2 plume between injection and production wells. For constant injection rate cases, they observed the injection pressure peak before the commencement of CO_2 production. They reported that the reservoir pressure and energy extraction were very sensitive to the injection flow rate, injection temperature and reservoir permeability, and the CO_2 injection temperature had notable thermo-siphon effects. However, they considered a very high initial reservoir temperature and pressure of 225°C and 150–200 bar, respectively. The thermo-physical data show that the difference between the fluid mobilities of water and CO_2 is smaller at higher temperatures and pressures. Babaei (2019) simulated a 3D CPG system with an initial reservoir temperature of 80°C. His focus was to study the CO_2 injectivity and storage efficiency for open and closed mass and heat flux boundary conditions. It was found that the discrepancies in geothermal heat recovery and pressure buildup were less when the number of injection and production wells was greater.

In past studies (Randolph and Saar, 2010; Adams et al., 2015; Babaei, 2019), the higher mobility of CO_2 than brine/water was mainly highlighted as an advantage for geothermal energy extraction. For temperatures above 60°C, the specific heat capacity of CO_2 is notably smaller than that of water. This may be a demerit of CPG from high-enthalpy reservoirs. However, thermo-physical data reveals that for temperatures between 35°C and 55°C and pressures around 10 MPa, CO_2 has a significantly higher specific heat capacity than water (see Figure 1.1). It can boost energy production rates after CO_2 breakthrough in the production well at low to moderate reservoir temperatures. Additionally, density, mobility, and enthalpy of CO_2 and CO_2 solubility in water vary significantly in the temperature range of 30°C–80°C and the pressure range of 10–20 MPa (Hnědkovský et al., 1996; Duan and Sun, 2003; Mao et al., 2013; Lemmon et al., 2019), which are representative of shallow sedimentary and low- to moderate-enthalpy reservoirs.

We have simulated two-phase CO_2-brine flow with CO_2 solubility in brine, heat transfer, and dissolved CO_2 transport for a 2D vertical plane. Our simulation results clearly demonstrate that the higher CO_2 heat capacity between 35°C and 55°C boosts energy production rate after CO_2 breakthrough in the production well at a lower reservoir temperature. The consequences of the buoyancy and injection pressure-driven free-phase CO_2 plume spreading on energy production and net CO_2 mass injection for different reservoir temperatures are discussed. The previous studies focused only on energy production and ignored the solubility of CO_2 in brine, which is essential for determining the CO_2 dissolution trapping potential in reservoir brine during and after energy production for CCS projects. Post injection-production, CO_2 spreading,

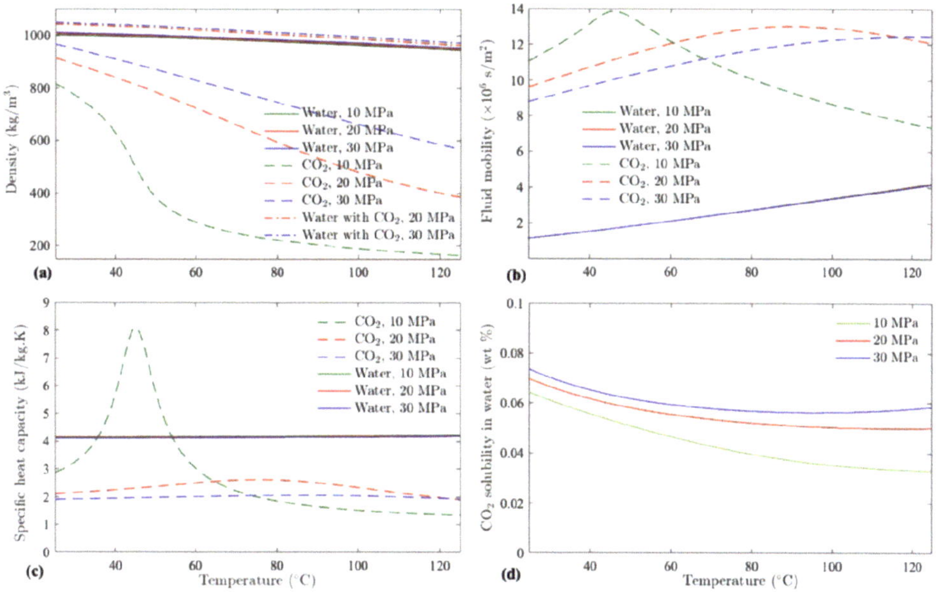

Figure 1.1 (a) Density comparison of fresh water, CO_2, and water with dissolved CO_2 (Hnědkovský et al., 1996; Duan and Sun, 2003; Mao et al., 2013; Lemmon et al., 2019), (b) fluid mobility of water and CO_2 (Lemmon et al., 2019), (c) specific heat capacity of water and CO_2, and (d) CO_2 solubility in water.

dissolution and trapping are also compared. We have shown the onset of buoyant convection in the diffusive boundary layer, which enhances the mixing of CO_2 in brine.

1.2 MATHEMATICAL MODELING

Simultaneous modeling of geothermal energy extraction and CO_2 dissolution trapping in a sedimentary reservoir by injecting liquid CO_2 and producing hot brine with CO_2 requires solving the following set of coupled mass balance equations, modified Darcy equation and heat transport equation for CO_2 and brine phases:

$$\frac{\partial}{\partial t}\left(\phi\sum_{\alpha}\rho_{\alpha}C_{\beta,\,\alpha}S_{\alpha}\right)+\nabla\cdot\left(\sum_{\alpha}\rho_{\alpha}C_{\beta,\,\alpha}\mathbf{q}_{\alpha}\right)-\nabla\cdot\left(\phi\sum_{\alpha}\rho_{\alpha}\mathbf{D}_{\beta,\,\alpha}S_{\alpha}C_{\beta,\,\alpha}\right)=0, \quad (1.1)$$

In the above equation, brine or wetting-phase fluid and supercritical/liquid CO_2 or non-wetting-phase fluids are represented by $\alpha = w$ and $\alpha = nw$, respectively. The components $\beta = W$ and $\beta = CO_2$ denote water and carbon dioxide, respectively. Reservoir permeability and porosity are denoted by k and ϕ, respectively. Here, density (kg/m^3), saturation, pressure (Pa), viscosity (Pa·s), and the relative permeability of the fluid phase α are shown by ρ_{α}, S_{α}, p_{α}, μ_{α}, and $k_{r,\,\alpha}$, respectively. The mass fraction of component β in fluid phase α is denoted as $C_{\alpha,\,\beta}$. The seepage flux $(\boldsymbol{q}_{\alpha})$ and dispersion coefficient $(\mathbf{D}_{\beta,\,\alpha})$ are given as follows:

$$q_\alpha = -\frac{kk_{r,\alpha}}{\mu_\alpha}(\nabla p_\alpha - \rho_\alpha \mathbf{g}) \tag{1.2}$$

$$\left(\mathbf{D}_{\beta,\alpha}\right)_{i,j} = \frac{(\alpha_L - \alpha_T)(q_\alpha)_i (q_\alpha)_j}{|q_\alpha|} + \alpha_T |q_\alpha| \delta_{ij} + D_m \delta_{ij} \tag{1.3}$$

The saturation and mass fractions satisfy the following closure relations:

$$S_w + S_{nw} = 1,\ C_{CO_2,w} + C_{W,w} = 1 \text{ and } C_{CO_2,nw} + C_{W,nw=1} \tag{1.4}$$

$$p_{nw} - p_w = p_{c0}\left(\hat{S}^{-1/\lambda} - 1\right)^{1-\lambda} \tag{1.5}$$

$$k_{r,w} = \left(1 - \left(1 - \hat{S}^{1/\lambda}\right)^{\lambda}\right)^2 \sqrt{\hat{S}} \quad \text{and} \quad k_{r,nw} = \left(1 - \hat{S}^{1/\lambda}\right)^{2\lambda} \sqrt{1 - \hat{S}} \tag{1.6}$$

In the above equation, $\hat{S}$ is the effective brine saturation. It is defined as $\hat{S} = (S_w - S_{rw})/(1 - S_{rw})$. S_{rw} is residual brine saturation. The capillary entry pressure is denoted by p_{c0}, and its magnitude is 10 kPa in this paper. $k_{r,w}$ and $k_{r,nw}$ are the relative permeabilities of brine and CO_2 phases, respectively. λ is the van Genuchten exponent value, and $\mathbf{g}$ is the gravitational acceleration vector ($-9.81\hat{k}$ m/s^2). Here, ϕ is porosity with magnitude 0.3, and k is the intrinsic reservoir permeability (10^{-13} m^2). The energy balance equation for multiphase fluid flow is given below:

$$\frac{\partial}{\partial t}\left((1-\phi)\rho_s C_{ps} T + \phi \sum_\alpha \rho_\alpha S_\alpha C_{p\alpha} T\right) + \nabla \cdot \left(\sum_\alpha \rho_\alpha \mathbf{q}_\alpha H_\alpha\right) - \nabla \cdot (\lambda_m \nabla T) = 0 \tag{1.7}$$

In the above equation, T (°C) is temperature, and C_{ps} and $C_{p\alpha}$ are the heat capacities of solid and fluid phases α, respectively. Here, H_α is the enthalpy of fluid phase α, and λ_m is the effective thermal conductivity of porous medium. Subscript s corresponds to the solid phase.

1.3 COMPUTATIONAL MODEL

Figure 1.2 presents the schematic diagram of a reservoir that is 1 km below the ground and spans 200 m vertically between caprock and baserock at the top and bottom, respectively. The injection well is placed in the middle of the reservoir (100 m below the caprock), whereas the production well is 10 m below the caprock. The horizontal distance between the injection and production wells is 1 km. Employing an injection well at a greater depth than the production well is advantageous for better pressure management inside the reservoir and widely suggested as a helpful design feature (Buscheck et al., 2013; Elliot et al., 2013; Garapati et al., 2014, 2015; Pan et al., 2018). Recent studies (Singh et al., 2019, 2020, 2021) on CO_2 sequestration in sedimentary reservoirs have shown that CO_2 rises above the injection well due to buoyancy and flows laterally below the caprock. This may cause early CO_2 production and affect energy production.

In the present study, CO_2 is injected at a fixed injection pressure and temperature (25°C) into reservoirs of three different temperatures: 75°C, 100°C and 125°C.

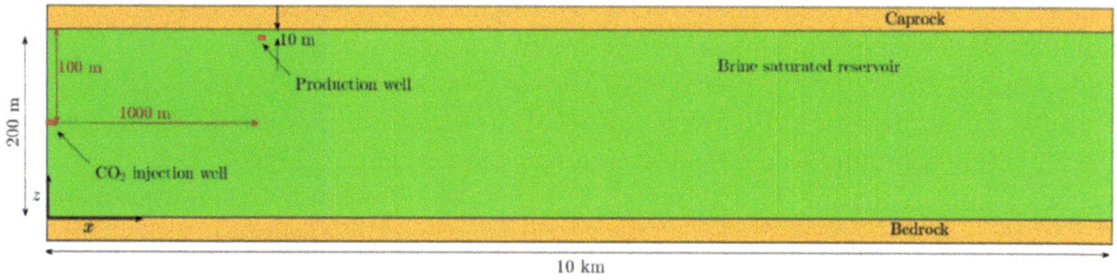

Figure 1.2 Schematic diagram of a reservoir for modeling CO_2-based geothermal energy production and CO_2 sequestration.

Table 1.1 Numerical modeling parameters used in this study

Parameter	*Value*
Reservoir temperature	75°C, 100°C, and 125°C
Reservoir pressure	10 MPa
Injection well pressure	13, 16, and 21 MPa
Production well pressure	9 MPa
Minimum residual water saturation	0.05
Minimum residual CO_2 saturation	0.05
Capillary entry pressure	10 kPa
Pore size distribution index	0.67
Rock density	2,560 kg/m^3
Rock-specific heat capacity	1,010 J/kg/°C
Thermal conductivity	1 W/m/°C

The reservoir permeability and porosity are 10^{-13}m^2 and 0.3, respectively. The reservoir was initially fully saturated with brine. In this study, the effect of dissolved salts in brine is not modeled, and the values of the thermo-physical properties of brine are taken as those of water without any dissolved salts. The simulation domain involves one injection well in between two production wells. Hence, a symmetric boundary condition has been imposed at the left boundary. At the top and bottom boundaries of the permeable reservoir, no flow, zero diffusive heat flux, and zero dispersive dissolved CO_2 flux are assigned as boundary conditions. At the far-field right boundary (in Figure 1.2), initial hydrostatic pressure, constant temperature, and a zero gradient for CO_2 saturation and dissolved concentration are imposed. The values of fluid properties are from NIST (Lemmon et al., 2019), and CO_2 solubility in water was modeled as suggested by Mao et al. (2013). The temperature- and pressure-dependent density, mobility, and specific heat capacity of different fluid phases are shown Figure 1.1a–c. Significant variations of CO_2 density, mobility, and specific heat capacity in the temperature range (30°C–80°C) and pressure range (10–20 MPa) are evident. Water density, mobility, and specific heat capacity vary mildly with temperature but negligibly with pressure. The variation of CO_2 solubility in water with temperature and pressure is shown in Figure 1.1d. We have used van Genuchten relative permeability and capillary pressure-saturation constitutive functions where the capillary entry pressure, exponent, and minimum residual brine saturation are 10 kPa, 0.67, and 0.05, respectively. The values of different parameters and constants used in this study are given in Table 1.1.

We have used FEHM (Finite Element Heat and Mass Transfer code) to simulate the geothermal heat extraction and CO_2 dissolution processes. FEHM is a control volume finite element method (CVFE)-based code (Zyvoloski et al., 2011). Using the fully implicit formulation, FEHM achieves system stability. FEHM uses two unique features for the functional representation of the equation of state (EOS) for fluids. It uses a rational polynomial EOS for water and water vapor. Due to this, the rational function approximation and tabular EOS calculate the analytical derivative during the Newton-Raphson iteration. These functions accurately represent the water properties and give the derivatives at almost no computational cost due to the polynomial character of the function. A third-order polynomial of pressure and temperature for both the numerator and denominator is used for density calculation (Zyvoloski, 2007). Adaptive time step-size was used, and relative tolerance was set at 10^{-6}. We used structured grids for the numerical discretization of two-phase flow, heat, and solute transport equations. The dimension of the grids in the z direction was uniform with $\Delta z = 2$ m. In x direction, the dimension of the grid was variable, with finer meshes near the wells and coarser meshes away from the wells. Near the wells, we used grids with $\Delta x = 2$m and $\Delta z = 2$m. The grid size near the well was chosen based on our prior grid sensitivity studies (Singh et al., 2019, 2021). It was shown that $\Delta x = 2$m and $\Delta z = 2$m were sufficient to accurately model the saturation front movement in the vicinity of the well, but final grid with $\Delta x = 1$m and $\Delta z = 1$m might be needed when the development of solutal fingering was modeled. In this study, the values of Δx were gradually increased to 40 m at locations 500 m away from the wells. The computational mesh is shown in Figure 1.3a. The total number of nodes is 246 and 101 in x and z directions, respectively. The computational time for simulating the coupled problem using FEHM varied between 0.4 and 1 month for different cases. However, a finer grid-based simulation was performed. A finer computational grid where Δx was increased to 20 m is shown in Figure 1.3b. For the finer computational grid, Δz and Δx near the wells were kept the same as 2 m. For 100°C reservoir temperature and 16 MPa injection pressure, the geothermal energy extraction by CO_2 was simulated using the computational grid shown in Figure

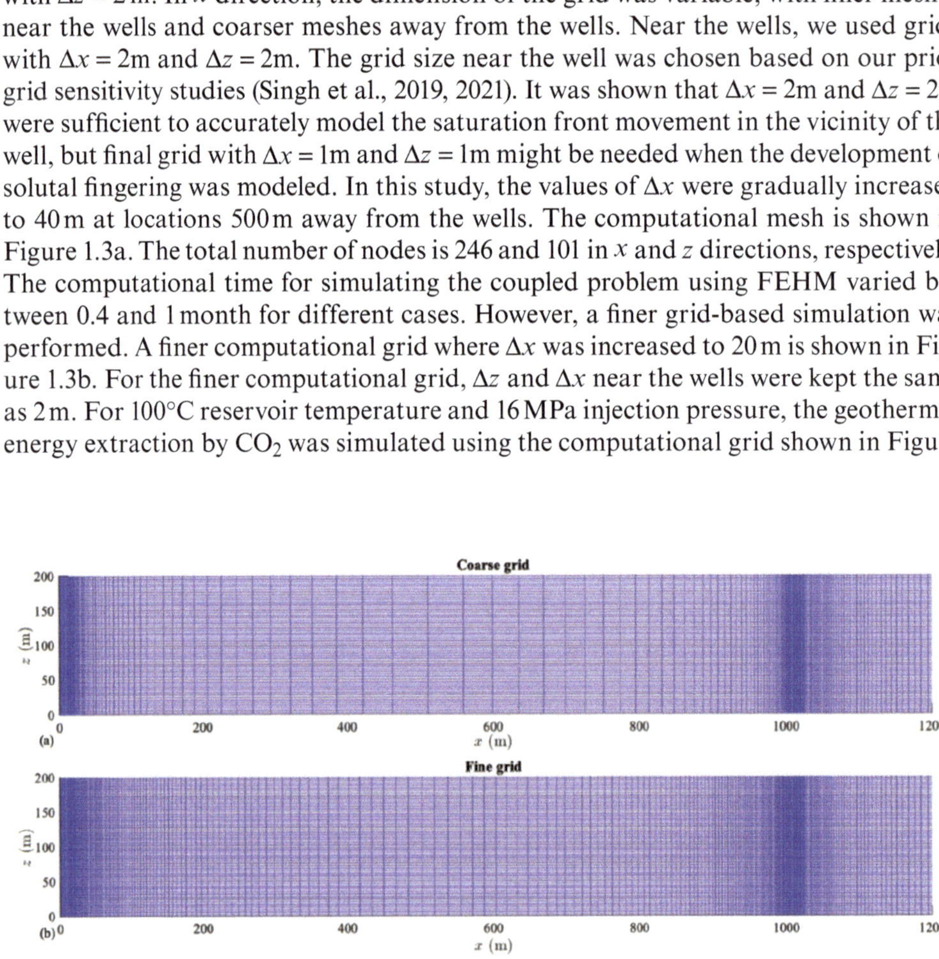

Figure 1.3 (a) Coarse grid and (b) fine grid used to investigate the grid sensitivity study.

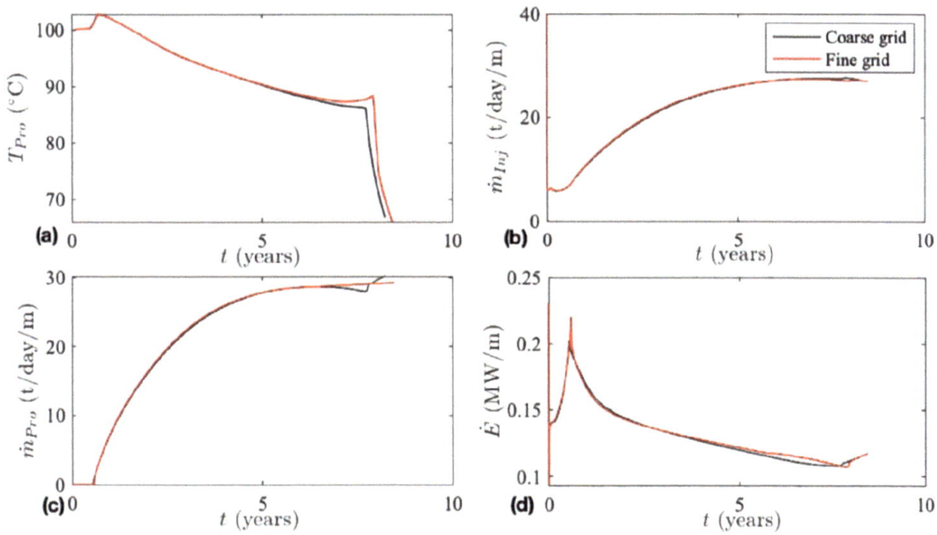

Figure 1.4 Comparison of (a) production temperature drawdown, (b) CO_2 injection rate, (c) CO_2 production rate, and (d) the overall energy production for coarser and finer grids.

1.3b. Four key output variables: production temperature drawdown, CO_2 injection rate, CO_2 production rate, and the overall energy production rate are compared in Figure 1.4 as obtained from coarse and fine grid simulations. Very minor differences in the results are observed. This demonstrates that the coarser computational mesh as described above is sufficient to study the energy production rate, temperature drawdown, and CO_2 injection rate for different reservoir temperatures and injection pressures.

1.4 RESULTS AND DISCUSSION

The numerical simulations of the CPG system were performed until the production well temperature (T_{pro}) reached 2/3rd of the initial reservoir temperature (T_r, °C). The duration of CO_2 injection and energy production is denoted as t_{CPG}. The simulations were then continued for a longer time to model the CO_2 plume migration in the reservoir and CO_2 dissolution trapping after the CPG system lifetime. The results associated with geothermal energy extraction by the CPG system are discussed in Section 1.4.1. Results pertaining to CO_2 plume migration and dissolution trapping in the post CPG operation phase are discussed in Section 1.4.2.

1.4.1 CO_2 injection and energy production period

The total energy produced (E) and net CO_2 injected (m_{CO_2}) before T_{pro} drops to 2/3rd T_r (°C) are plotted against injection pressure (P_I) in Figure 1.5a for nine different combinations of reservoir temperature and injection pressure. Table 1.2 presents the

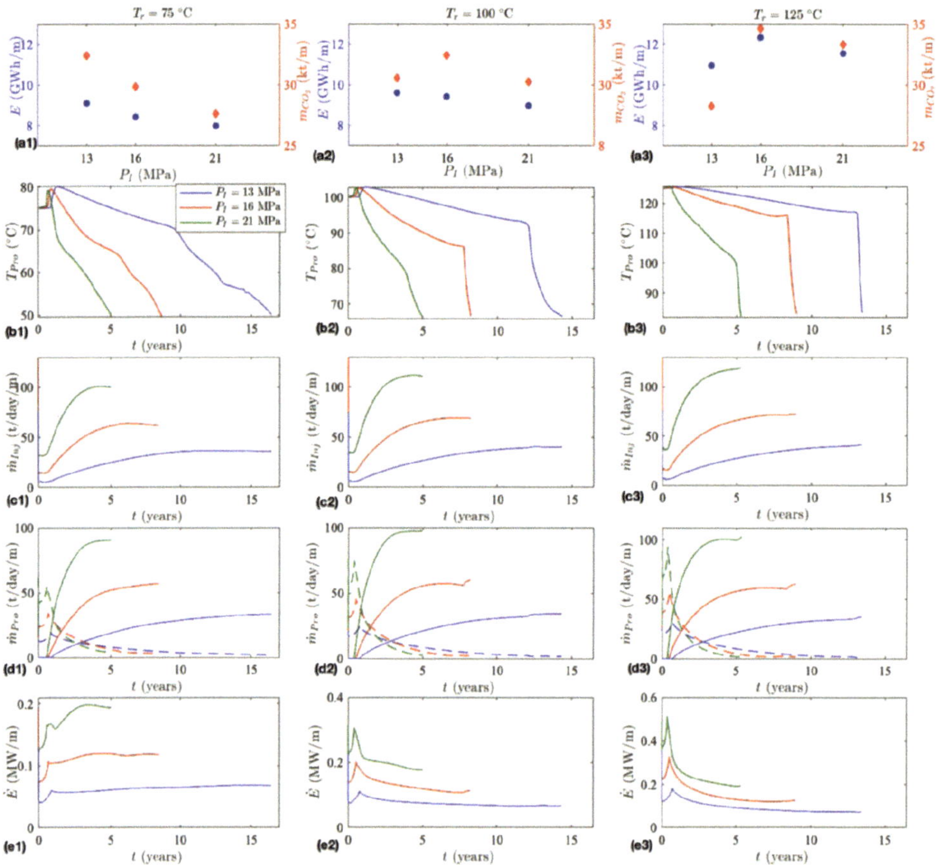

Figure 1.5 Comparison of (a) total energy produced $\left(E = \int_0^{t_{CPG}} \dot{E}\, dt\right)$ and net mass of injected CO_2 $m_{CO_2} = \int_0^{t_{CPG}} \left(\dot{m}_{pro,\, CO_2} - \dot{m}_{inj,\, CO_2}\right) dt$, (b) production well temperature (T_{pro}) drawdown curves, (c) injection rate of CO_2 ($\dot{m}_{inj,\, CO_2}$), (d) production rates of CO_2 (solid line, ($\dot{m}_{pro,\, CO_2}$)) and brine (dash line, ($\dot{m}_{pro,\, brine}$)), and (e) energy production rate ($\dot{E} = \dot{m}_{pro,\, CO_2} H_{pro,\, CO_2} + \dot{m}_{pro,\, brine} H_{pro,\, bri-e} - \dot{m}_{inj,\, CO_2} H_{inj,\, CO_2}$), where $H_{inj,\, CO_2}$, $H_{pro,\, CO_2}$, and $H_{pro,\, brine}$ are the enthalpy of injected CO_2, produced CO_2 and produced brine. The injection and production results are for unit width of the reservoir. CO_2 is injected at a constant temperature (25°C) and pressure, while the production well pressure is specified as 9 MPa.

Table 1.2 The comparison of time (t_{CPG}) of production temperature dropping to 2/3rd of T_r (°C), total extracted energy and the total mass of remaining CO_2 during the geothermal energy production till t_{CPG}

T_r (°C)	P_I (MPa)	t_{CPG} (days)	E (GJ/m)	m_{CO_2} (kt/m)
75	13	6,000.8	32,690	32.42
	16	3,179.6	30,290	29.86
	21	1,841.8	28,670	27.61
100	13	5,236.0	34,600	30.56
	16	2,998.8	33,900	32.44
	21	1,823.7	32,400	30.28
125	13	4,893.9	39,500	28.28
	16	3,278.4	44,460	34.63
	21	1,922.7	41,567	33.35

same information in more detail. However, the values of E and m_{CO_2} vary significantly with injection pressure, while reservoir temperature is the same. Figure 1.5a1 shows that for T_r = 75°C, both E and m_{CO_2} increase as P_I decreases from 21 to 13 MPa. But Figure 1.5a3 shows that for T_r = 125°C both E and m_{CO_2} are the highest when P_I = 16 MPa. The results in Figure 1.5a2 for T_r = 100°C show the trends of variation of E and m_{CO_2} variation with P_I are consistent with those shown in Figure 1.5a1 and a3 for T_r = 75°C and 125°C. The surprising trends in Figure 1.5a are the result of large variations in the mobility, density, and heat capacity of CO_2 with pressure and temperature.

The temperature drawdown curves for nine cases are shown in Figure 1.5b. The corresponding liquid CO_2 mass injection rates ($\dot{m}_{CO_2,inj}$), brine and liquid/supercritical CO_2 mass production rates ($\dot{m}_{B,\ pro}$ and $\dot{m}_{CO_2,pro}$), and energy production rates ($\dot{E}$) are plotted in Figure 1.5c–e. Figure 1.5b shows that for T_r = 75°C, the production temperature decreases gradually, but for T_r = 100°C and 125°C, the drawdown curves exhibit a sharp decline in T_{pro} after a certain period. The production temperature drawdown curves in Figure 1.5b show a stronger Joule-Thompson (JT) effect due to pressure drop at the production well (Stauffer et al., 2014). The temperature rise due to the JT effect is prominent at lower reservoir temperatures. Figure 1.5c shows that the CO_2 injection rate increases with time. The time when the CO_2 injection rate starts increasing matches the time of onset of free-phase CO_2 production, as shown in Figure 1.5d. Before the onset of CO_2 production, brine production rates increase. There may be two reasons behind the increase in total mass of fluid (brine and free-phase CO_2 together) production rate in Figure 1.5d: (i) thermo-siphon effect (Buscheck et al., 2013; Elliot et al., 2013; Pan et al., 2018) due to expansion of CO_2 inside the reservoir at lower pressure and higher temperature, and (ii) the increase of effective mobility with the increase of less viscous CO_2 saturation inside the reservoir (Singh et al., 2019) causing higher fluid flow under the same pressure difference between injection and production wells. Singh et al. (2019) found decreasing injection pressure with time for a constant CO_2 injection rate in a sedimentary storage reservoir. The energy production rate, which depends on the mass outflow rate of brine and free-phase CO_2 and their temperature at the production well, is seen in Figure 1.5e to increase during the first few months after the commencement of CPG operation. At $T_r = 75°C$, the energy production rate does not decrease much in time, as seen in Figure 1.5e1, because the heat capacity of CO_2 is higher at lower temperatures (between 35°C and 55°C) when the pressure is close to 10 MPa. However, for $T_r = 100°C$ and 125°C, a steady decrease in energy production rate after CO_2 breakthrough is observed in Figure 1.5e2 and e3.

The migration of free-phase CO_2 (in a liquid or supercritical) plume, cooling of the reservoir and CO_2 dissolution during the injection-production period are illustrated

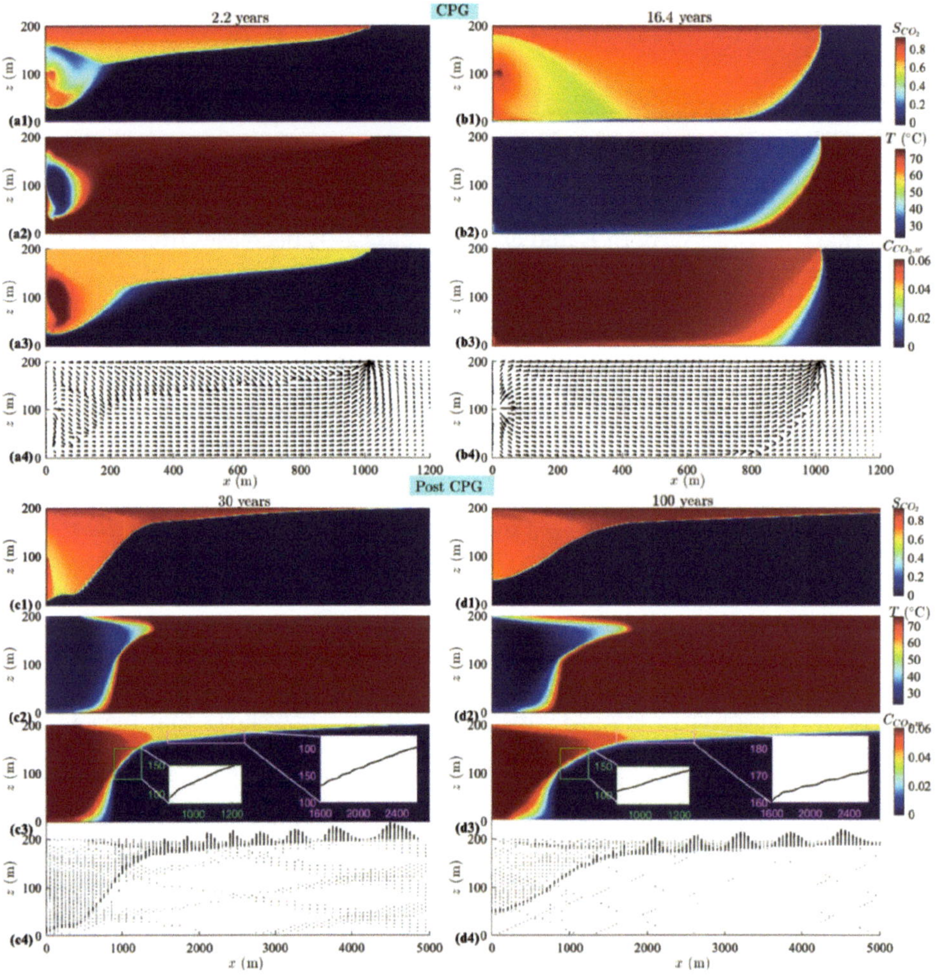

Figure 1.6 Temporal evolution of field variables for 100°C reservoir temperature and 16 MPa injection pressure: snapshots of (a1, b1, c1, d1) free-phase CO_2 saturation distribution, (a2, b2, c2, d2) temperature distribution, (a3, b3, c3, d3) dissolved CO_2 concentration distribution, and (a4, b4, c4, d4) total velocity vectors at (a) 1.1, (b) 8.2 years during injection-production, and (c) 30 and (d) 100 years after injection-production.

in Figure 1.6a and b by plotting CO_2 saturation (S_{CO_2}), temperature (T), dissolved CO_2 concentration in brine ($C_{CO_2,w}$) and total velocity vectors at 1.1 and 8.2 years for $T_r = 100°C$, $P_I = 16$ MPa. In this case, the production temperature (T_{pro}) becomes 2/3rd of T_r after 8.2 years. Free-phase CO_2 saturation plots in Figure 1.6a1 and b1 depicts that buoyant free-phase CO_2 rises upward from the injection well and then migrates laterally toward the production well. After the tip of the free-phase CO_2 plume reaches the production well, the CO_2 saturation front moves downward, as understood from Figure 1.6b1. Figure 1.6a4 and b4 show sharp changes of velocity magnitude and direction right below the free-phase CO_2 plume, i.e., the capillary transition zone (CTZ). At early times, the outward velocity vectors at the edge of CTZ in Figure 1.6a4 imply the downward and rightward movement of the free-phase CO_2 saturation front. The velocity vector plots outside CTZ corresponds to brine flow which is very small when compared to the CO_2 flow inside the CTZ. The heat depletion front moves slower than the free-phase CO_2 saturation front. This is evident from Figure 1.6a2. But later, the cold region closely matches the CTZ, as seen in Figure 1.6b2 and d2. Corresponding dissolved CO_2 concentration distributions, which are shown in Figure 1.6a3 and b3 closely follow the CTZ, but higher concentrations are observed in the cold regions because CO_2 solubility in water/brine decreases with temperature in the present temperature and pressure ranges. Outside CTZ, the thin diffusive boundary layer of CO_2 concentration is seen. The CO_2 plume migration and cooling of the reservoir at the time of $T_{pro} = 2/3T_r$ (°C) for $P_I = 13$ and 21 MPa, and 100°C reservoir temperature are shown in Figure 1.7. The differences in the distributions of S_{CO_2} and T at different injection pressures are apparent. The variations of E and m_{CO_2} with P_I are linked with the spatial variation of S_{CO_2} and T. The variations of S_{CO_2}, T, $C_{CO_2,w}$, and total velocity vectors at 2.2 and 16.4 years for $T_r = 75°C$, $P_I = 13$ MPa are shown in Figure 1.8.

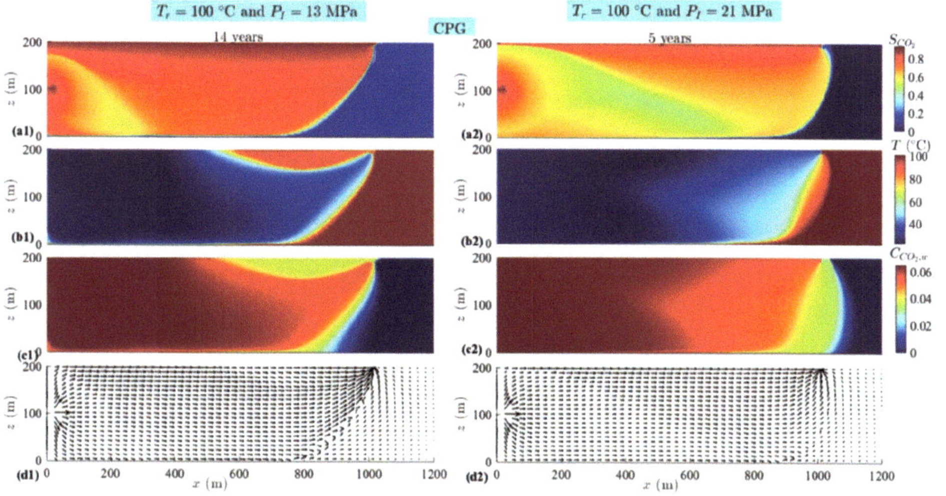

Figure 1.7 Comparison of plume patterns, temperature distribution between two injection pressures (a) 13 MPa and (b) 21 MPa for 100°C reservoir temperature at the end of CO_2 injection and energy production, i.e., $T_{pro} = 2/3T_r$ (°C).

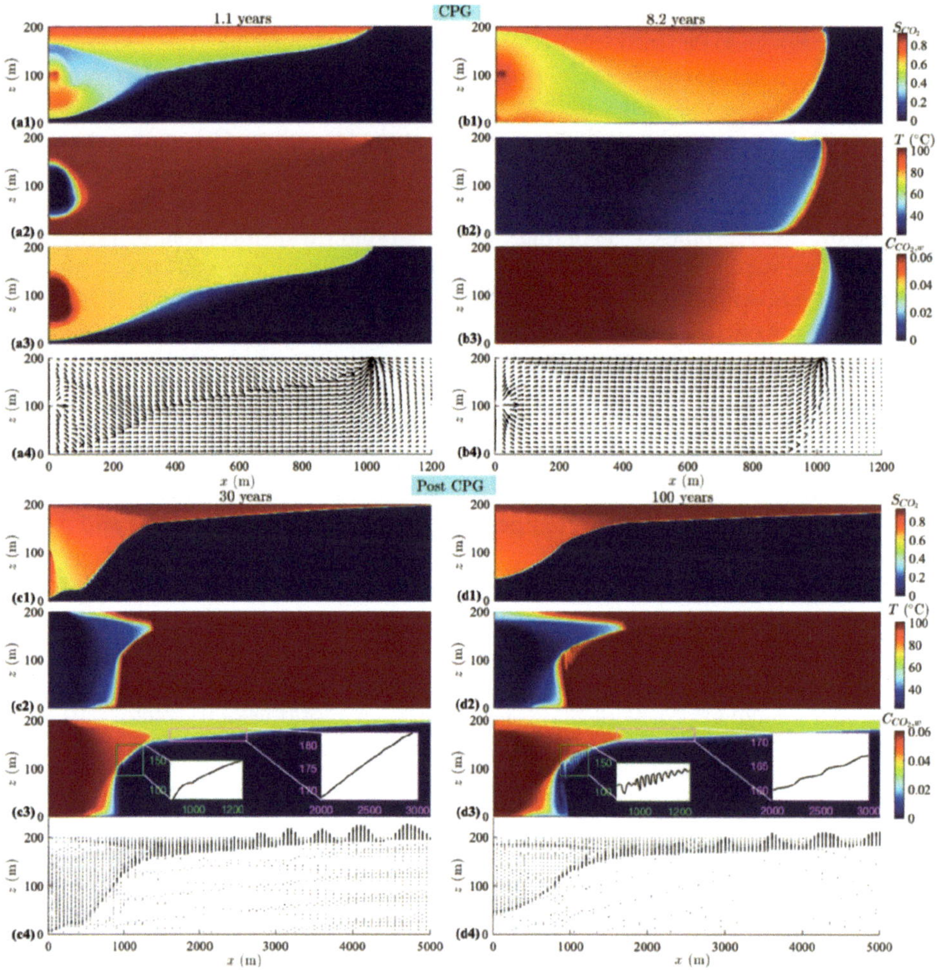

Figure 1.8 Temporal evolution of field variables for 75°C reservoir temperature and 13 MPa injection pressure: snapshots of (a1, b1, c1, d1) free-phase CO_2 saturation distribution, (a2, b2, c2, d2) temperature distribution, (a3, b3, c3, d3) dissolved CO_2 concentration distribution, and (a4, b4, c4, d4) total velocity vectors at (a) 2.2, (b) 16.4 years during injection-production, and (c) 30 and (d) 100 years after injection-production.

1.4.2 Post injection-production period

Post CPG, i.e., after CO_2 injection and energy production, the free-phase CO_2 continues spreading below the caprock and dissolves in brine. Figure 1.9a shows the change of plume tip position (x_{tip}) with time, which is measured from the end of CO_2 injection and energy production. Free-phase CO_2 plume velocity in the post injection-production period takes 20%–30% less time for $T_r = 125°C$ than $T_r = 75°C$ when P_I is the same. It is because total mobility at the saturation front ($k_{r,w}\mu_w + k_{r,nw}\mu_{nw}$) increases with temperature. Here $k_{r,w}$ and $k_{r,nw}$ are the relative permeabilities, while μ_w and μ_{nw} are

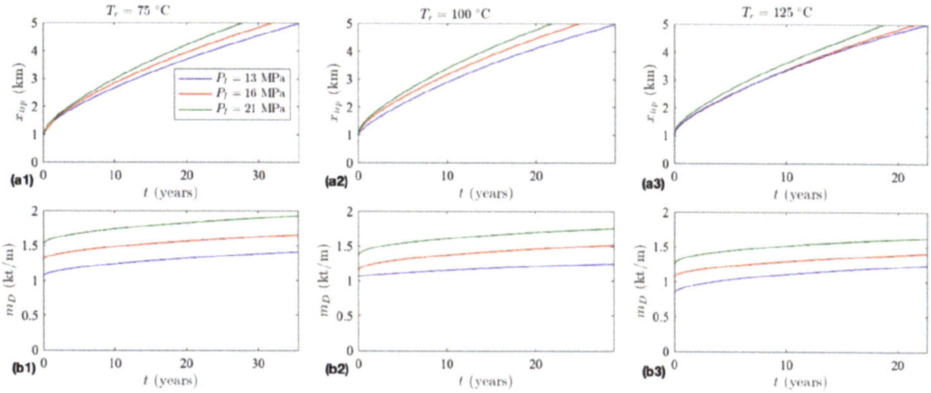

Figure 1.9 Comparison of (a) free-phase CO_2 plume tip migration and (b) total dissolved CO_2 in the reservoir for the entire injection and post-injection duration times.

the viscosities of water and CO_2, respectively. The comparison of dissolved CO_2 mass during and post CPG is important to know which conditions favor better dissolution trapping. The values of m_D at $t = 0$ in Figure 1.9b1–b3 corresponds to the total mass of CO_2 dissolved during CPG operation. At the end of CPG, a large fraction of CO_2 inside the reservoir is in the free phase. Figure 1.9b1–b3 shows that for lower reservoir temperatures, the mass of dissolved CO_2 is higher because of higher solubility at lower temperatures when the pressure range is between 10 and 20 MPa (see Figure 1.2e). For higher injection pressure, m_D at the end of CPG is more because CO_2 solubility also increases with pressure. Post CPG, the increase in contact area between brine and free-phase CO_2 due to free-phase CO_2 plume spreading, diffusion, and convection results in an increase in dissolved CO_2 mass.

The free-phase CO_2 plume evolution during post CPG is shown in Figure 1.6c1 and d1 for $T_r = 100$°C, $P_I = 16$ MPa. The CO_2, which flew below the injection well during CPG due to injection pressure, rises and laterally migrates further in the rightward direction. Figure 1.6c2 and d2 shows that the cooling front within 25–50 m below the top boundary moves rightwards, but adjacent to the top boundary, the cooling front is seen to retreat. Figure 1.6c3 and d3 shows higher concentrations in the cold region of the free-phase CO_2 plume and lower concentrations in the triangular-shaped region of the free-phase CO_2 plume. In Figure 1.6d3, solutal fingers are prominent. In the inset figures of Figure 1.6c3 and d3, the wavy nature of the concentration contour lines for $C_{CO_2, w} = 0.02$ wt% depicts gravitational instability and the development of solutal fingers at the flat portion of the diffusive boundary layer. Such solutal fingers are not observed in the case of a 75°C reservoir temperature (see Figure 1.8). This implies faster formation of solutal fingering at a 100°C reservoir temperature. The fluctuating vertical velocity at the far-field interface between free-phase CO_2 and brine, as shown in Figure 1.6c4 and d4 also suggests the onset of buoyant convection due to gravitational instability at the thin diffusive boundary layer below the capillary transition zone. Here, we would like to mention that to apprehend the onset of solutal fingering and simulate their growth, finer grids are necessary, and it was not possible for the present setup due to the enormous computational cost.

1.5 CONCLUSIONS

Uncertainties in the energy markets across the globe and developing concerns due to climate change are pushing industries and governments toward greener or less carbon-intensive energy resources. Using carbon dioxide to extract geothermal energy from hot and deep sedimentary reservoirs is one such combination that limits the anthropogenic emission of CO_2 in the atmosphere to avert greenhouse gas-driven climate change as well as ensures a continuous supply of low-carbon energy from the earth's crust. Therefore, CO_2 plume-based geothermal energy extraction from reservoirs of low to moderate enthalpy is investigated based on numerical simulations. The effects of reservoir temperatures and injection pressures on the net amount of injected CO_2 into the reservoir during CPG and post CPG free-phase CO_2 spreading and dissolution in brine are also studied.

We have found that for a constant injection pressure case, the injection rate of CO_2 increases with time once it reaches the production well. This influences the energy production rate. Even though the energy production rate from lower reservoir temperatures is initially smaller, it increases once CO_2 production starts due to the higher heat capacity of CO_2 in the temperature range of 35°C–55°C. For T_r = 100°C and 125°C, the energy production rate steadily decreases after free-phase CO_2 production starts. This is an important highlight of the present numerical study.

The simulation results for different injection pressures suggest that for a lower reservoir temperature (T_r = 75°C), a smaller injection pressure results in higher total energy recovery and a higher net mass of CO_2 sequestration before the production temperature drops to 2/3rd of the initial reservoir temperature. But for higher reservoir temperatures, intermediate injection pressure may be beneficial for both total energy extraction and net CO_2 sequestration. Hence, high CO_2 injection pressure may not be advisable when a CO_2-based EGS is designed for maximum energy recovery from a low-enthalpy geothermal reservoir. High injection pressure also has adverse effects on the structural integrity of the caprock.

While the free-phase CO_2 migrates further below the caprock during the post-injection period, CO_2 slowly dissolves in the brine. For smaller reservoir temperatures, CO_2 plume migration is slower, but the dissolution rate is higher because CO_2 solubility is a decreasing function of temperature. The CO_2 dissolution also triggers gravitational instability and enhances the dissolution trapping.

ACKNOWLEDGEMENTS

The authors gratefully acknowledge the computational facility of PG Senapathy Computer Center, IIT Madras. This work is part of the Ph.D. thesis of Mrityunjay Singh at IIT Madras, funded by MHRD, Government of India. The authors thank Philip H. Stauffer for his support in FEHM. The FEHM executable file used for simulations is freely available at the LANL website: https://fehm.lanl.gov/downloads/index.shtml. The input file for simulation using FEHM and some output files are publicly available in Mendeley dataset: https://data.mendeley.com/datasets/s8vp2wm8p6/1.

REFERENCES

Adams, B. M., Kuehn, T. H., Bielicki, J. M., Randolph, J. B., and Saar, M. O. (2015). A comparison of electric power output of CO_2 Plume Geothermal (CPG) and brine geothermal systems for varying reservoir conditions. *Applied Energy*, 140:365–377.

Babaei, M. (2019). Integrated carbon sequestration–geothermal heat recovery: Performance comparison between open and close systems. *Transport in Porous Media*, 126(1):249–273.

Brugger, J., Long, N., McPhail, D., and Plimer, I. (2005). An active amagmatic hydrothermal system: The paralana hot springs, northern flinders ranges, South Australia. *Chemical Geology*, 222(1):35–64.

Buscheck, T. A., Elliot, T. R., Celia, M. A., Chen, M., Sun, Y., Hao, Y., Lu, C., Wolery, T. J., and Aines, R. D. (2013). Integrated geothermal-CO_2 reservoir systems: Reducing carbon intensity through sustainable energy production and secure CO_2 storage. *Energy Procedia*, 37:6587–6594.

Davidson, C., Heldebrant, D., Bearden, M., Horner, J., and Freeman, C. (2017). Enabling ccs via low-temperature geothermal energy integration for fossil-fired power generation. *Energy Procedia*, 114:6448–6454.

Duan, Z. and Sun, R. (2003). An improved model calculating CO_2 solubility in pure water and aqueous NaCl solutions from 273 to 533 K and from 0 to 2000 bar. *Chemical Geology*, 193(3–4):257–271.

Elliot, T., Buscheck, T., and Celia, M. (2013). Active CO_2 reservoir management for sustainable geothermal energy extraction and reduced leakage. *Greenhouse Gases: Science and Technology*, 3:50–65.

Garapati, N., Adams, B. M., Bielicki, J. M., Schaedle, P., Randolph, J. B., Kuehn, T. H., and Saar, M. O. (2017). A hybrid geothermal energy conversion technology – A potential solution for production of electricity from shallow geothermal resources. *Energy Procedia*, 114:7107–7117.

Garapati, N., Randolph, J. B., and Saar, M. O. (2015). Brine displacement by CO_2, energy extraction rates, and lifespan of a CO_2-limited CO_2-Plume Geothermal (CPG) system with a horizontal production well. *Geothermics*, 55:182–194.

Garapati, N., Randolph, J. B., Valencia, J. L., and Saar, M. O. (2014). CO_2-plume geothermal (CPG) heat extraction in multi-layered geologic reservoirs. *Energy Procedia*, 63:7631–7643.

Gosnold, W. D. (2015). *Electric Power Generation from Low to Intermediate Temperature Resources*. Report published by: University of North Dakota, United States: N. p. Web. doi:10.2172/1186077

Hnědkovský, L., Wood, R. H., and Majer, V. (1996). Volumes of aqueous solutions of CH_4, CO_2, H_2S and NH_3 at temperatures from 298.15 k to 705 k and pressures to 35 MPa. *The Journal of Chemical Thermodynamics*, 28(2):125–142.

Lemmon, E., McLinden, M., and Friend, D. (2019). Thermophysical properties of fluid systems, in *NIST Chemistry Webbook, NIST Standard Reference Database Number 69*, (20899). P.J. Linstrom and W.G. Mallard, eds., National Institute of Standards and Technology, Gaithersburg, MD, https://doi.org/10.18434/T4D303.

Mao, S., Zhang, D., Li, Y., and Liu, N. (2013). An improved model for calculating CO_2 solubility in aqueous NaCl solutions and the application to $CO_2 - H_2O$ – NaCl fluid inclusions. *Chemical Geology*, 347(May):43–58.

Mohan, A. R., Turaga, U., Subbaraman, V., Shembekar, V., Elsworth, D., and Pisupati, S. V. (2015). Modeling the CO_2-based enhanced geothermal system (EGS) paired with integrated gasification combined cycle (IGCC) for symbiotic integration of carbon dioxide sequestration with geothermal heat utilization. *International Journal of Greenhouse Gas Control*, 32:197–212.

Pan, S.-Y., Gao, M., Shah, K. J., Zheng, J., Pei, S.-L., and Chiang, P.-C. (2019). Establishment of enhanced geothermal energy utilization plans: Barriers and strategies. *Renewable Energy*, 132:19–32.

Pan, C., Romero, C. E., Levy, E. K., Wang, X., Rubio-Maya, C., and Pan, L. (2018). Fully coupled wellbore-reservoir simulation of supercritical CO_2 injection from fossil fuel power plant for heat mining from geothermal reservoirs. *Journal of CO_2 Utilization*, 27(September):480–492.

Randolph, J. B. and Saar, M. O. (2010). Coupling geothermal energy capture with carbon dioxide sequestration in naturally permeable, porous geologic formations: A comparison with enhanced geothermal systems. *GRC Transactions*, 34:433–438.

Randolph, J. B. and Saar, M. O. (2011). Combining geothermal energy capture with geologic carbon dioxide sequestration. *Geophysical Research Letters*, 38(10):L10401.

Reed, M. J. (1983). Assessment of low-temperature geothermal resources of the United States-1982. Geological Survey Circular 892, (Published by Department of Interior: United-States Geological Survey). https://doi.org/10.3133/cir892

Singh, M., Chaudhuri, A., Chu, S. P., Stauffer, P. H., and Pawar, R. J. (2019). Analysis of evolving capillary transition, gravitational fingering, and dissolution trapping of CO_2 in deep saline aquifers during continuous injection of supercritical CO_2. *International Journal of Greenhouse Gas Control*, 82:281–297.

Singh, M., Chaudhuri, A., Soltanian, M. R., and Stauffer, P. H. (2021). Coupled multiphase flow and transport simulation to model CO_2 dissolution and local capillary trapping in permeability and capillary heterogeneous reservoir. *International Journal of Greenhouse Gas Control*, 108:103329.

Singh, M., Chaudhuri, A., Stauffer, P. H., and Pawar, R. J. (2020). Simulation of gravitational instability and thermo-solutal convection during the dissolution of co in deep storage reservoirs. *Water Resources Research*, 56(1):e2019WR026126. doi: 10.1029/2019WR026126.

Stauffer, P. H., Lewis, C., Stein, J., Travis, B., Lichtner, P., and Zyvoloski, G. (2014). Joule-Thomson effects on the flow of liquid water. *Transport in Porous Media*, 105:471–485.

Thorsteinsson, H., Augustine, C., Anderson, B. J., Moore, M. C. M. C., Tester, J. W., Hildigunnur, T., Augustine, C., Anderson, B. J., Moore, M. C. M. C., and Tester, J. W. (2008). The impacts of drilling and reservoir technology advances on EGS. *Proceedings, Thirty-Third Workshop on Geothermal Reservoir Engineering*. Stanford University, Stanford, CA, January 28–30, 2008SGP-TR-185.

Yadav, K. and Sircar, A. (2020). Geothermal energy provinces in India: A renewable heritage. *International Journal of Geoheritage and Parks*, 9:93–107.

Zyvoloski, G. A. (2007). *FEHM: A Control Volume Finite Element Code for Simulating Subsurface Multiphase Multi-Fluid Heat and Mass Transfer*. Los Alamos National Laboratory, New Mexico.

Zyvoloski, G. A., Robinson, B., Dash, Z., and Trease, L. (2011). *Summary of the Models and Methods for the FEHM Application – A Finite-Element Heat- and Mass-Transfer Code*. Los Alamos National Laboratory, Los Almos, New Mexico. LA-13306-MS.

Chapter 2

EGS development in Europe based on operational geothermal projects

Albert Genter, Clément Baujard, Eléonore Dalmais, Carole Glaas, Vincent Maurer, and Guillaume Ravier
ES-Géothermie

CONTENTS

2.1 INTRODUCTION: HOW TO ENHANCE THE PERMEABILITY?

Deep geothermal energy needs two main intrinsic factors to be largely used: a sufficient temperature and a sufficient flow rate for achieving an economically viable project. As it is not possible to increase temperature, scientists and engineers imagined producing more flow rate by introducing the concept of enhanced geothermal systems (EGS) also known as engineered geothermal systems. Therefore, EGS appears to be the way to develop geothermal energy in geographical areas reputed to be depleted in conventional hydrothermal resources.

As we cannot increase the temperature of the resource except by doing heat storage, it looked possible for engineers to increase the flow rate (injectivity or productivity) of a given borehole by applying innovative techniques that enhance the hydraulic performance. These techniques are called THMC stimulations (Thermal, Hydraulic, Mechanical and Chemical).

Thermal stimulation, which is typically performed to enhance the near-well field permeability, consists of injecting cold and thus dense water within a hot geological formation. The expected result is the development of thermal stresses, which generally

DOI: 10.1201/9781003271475-2

provoke a localized contraction of the rock mass at the borehole wall and thus the creation of micro- to mesofractures able to enhance the flow. Thermoelastic effects reducing the normal stress on fractures and thus promoting fracture opening could be considered (Dempsey et al., 2014).

Hydraulic (or Mechanical) stimulation is a more conventional technique because it is largely used in the oil and gas industry. It consists of injecting a significant volume of water into the rock mass at a relatively high overpressure. The mechanical process targeted to improve permeability is generally not fracking (i.e., creating new fractures in the rock mass) but shearing of pre-existing fractures. The fracture shearing process, which creates sufficient pathways to increase the post-stimulation flow rate, is associated with induced seismicity. In EGS, during hydraulic stimulation, or hydro-shearing, overpressure induces slip along pre-existing fractures that are favourably oriented in the stress field for reactivation in shear. In the case of the Upper Rhine Graben (URG), such hydraulic stimulation is usually performed over a large open-hole section of the borehole.

Chemical stimulation consists generally of dissolving some rock-forming minerals at the borehole or at some distance from the borehole. Thus, chemicals are injected and could dissolve some selected minerals, allowing pathway enlargement for flow and thus a high post-stimulation flow rate. Generally, an acid fluid is injected to dissolve materials around the wellbore, such as cuttings from the drilling and hydrothermal minerals plugging the fracture zones. This technique, largely used in the oil and gas industry, was also adapted to EGS sites in the URG (Nami et al., 2008; Portier et al., 2009; Schindler et al., 2010; Vidal et al., 2016).

All the above-mentioned techniques compete to increase the production flow rate or lower the reinjection pressure, which means increasing the well injectivity. However, it has been observed that stimulation experiments generate induced seismicity, which helps to increase hydraulic performances (Schill et al., 2017; Baujard et al., 2017; Maurer et al., 2020) but with the main drawback of being occasionally felt on the surface, like in the Strasbourg area between late 2019 and mid-2021 with magnitudes below 4 (Schmittbuhl et al., 2021).

During geothermal plant operation, it has also been observed that induced seismicity could take place generally close to the reinjection well, where the reinjected temperature is colder and thus the geothermal water becomes denser, inducing a thermo-mechanical effect on the rock mass (Maurer et al., 2020). The positive achievement is that we never observed any felt seismicity during the exploitation of the two geothermal plants in Northern Alsace (France) at Soultz-sous-Forêts (SsF) and Rittershoffen over the last decade. It demonstrated that we could exploit EGS sites safely in Europe with no impact on the local population. This lesson is important to disseminate because the perception of geothermal energy is controversial due to an industrial accident observed in the Strasbourg area (Schmittbuhl et al., 2021).

2.2 EGS IN THE UPPER RHINE GRABEN

The main achievements of EGS projects in continental Europe were done in the URG, which is a Tertiary graben made with a Paleozoic basement overlaid by a thick Mesozoic and Cenozoic sedimentary pile. The early concept dealt with Hot Dry Rock (HDR) because vintage petroleum data highlighted temperatures higher than 100°C at less than 1 km deep in the sedimentary formations (Gérard and Kappelmeyer, 1987). It means that heat can be found at relatively shallow depths in hard rocks, limiting the economic impact of expansive drilling costs, which could represent between 40% and 50% of the CAPEX costs of a geothermal project in this area. Thus, many exploration research projects, like deep drilling to 5 km in a granitic basement, were conducted at SsF by a European scientific partnership involving mainly France, Germany, Switzerland, and the UK (Genter et al., 2010).

Over several decades of geothermal development in this area (Vidal and Genter, 2018), a series of geothermal plants are running today in France (SsF and Rittershoffen), Germany (Landau, Insheim, and Bruchsal) and Switzerland (Riehen). Those sites had penetrated the sedimentary formations of the graben and, for some of them, the deeper crystalline basement (Soultz, Rittershoffen, Insheim, Landau). All sites, except Riehen, are located inside the Rhine Graben (Figure 2.1).

In Soultz, Landau, or Insheim, thermal anomalies, which coincide with oil fields (Figure 2.1), are interpreted as the location of natural hydrothermal convective systems (Kohl et al., 2000). Natural hot fluids circulate very slowly, with a flux of a few decimetres per year, within large convection loops of several thousand metres. Fluids are channelled in the network of nearly vertical fractures and faults created by the tectonic history of the Rhine Graben. The conceptual model of the fractured reservoir corresponds to fractured zones or faults intersecting successively the top crystalline basement, the Permo-Triassic sandstones, and the Muschelkalk limestones, which constitute the hanging wall of the thermal cap rock (Figure 2.2) (Vidal and Genter, 2018). The matrix porosity is considered as negligible even within the sedimentary units.

There were very few deep geothermal development initiatives prior to that of SsF in the Rhine Graben. The first project is located in Cronenbourg, in the suburbs of Strasbourg, where the first geothermal well drilled in 1980, which targeted the Permo-Triassic sandstones at 3,000 m depth, did not achieve the expected flow rate due to the lack of natural permeability. In the case of Cronenbourg, the initial conceptual model reflected the exploitation of a matrix geothermal resource comparable to that of the Paris Basin, whereas it is probably the presence of fractures/faults carrying the initial permeability that should have been targeted. Then, it would have been improved hydraulically by applying a suitable hydro-chemical treatment, for example, but at that time stimulation techniques were poorly developed. Thus, this first deep geothermal project in the Strasbourg area was abandoned.

However, in the URG, it turns out that each time drilling operations penetrated the deepest crystalline basement reputed to be impermeable, significant permeability indicators (mud losses, gas anomalies, brine outflow, occurrences of specific secondary minerals, high rate of penetration, ...) were observed. For instance, Vuataz et al. (1990) observed a permeable zone related to a fractured zone in the granitic basement at SsF. Such a permeable zone is characterized by a severe gas anomaly bearing 550 ppm of helium dissolved in the geothermal brine at 1,820 m depth. Extensive structural analyses and hydrothermal alteration mineralogical studies in the granite from core data and cuttings samples have shown that fractured zones represent the main pathway for the geothermal fluids. A detailed fracture survey in the Triassic and Permian sandstones as well as in the granite basement showed that natural fractures systematically exhibit hydrothermal secondary deposits such as geodic quartz, clay minerals (illite, I/S, and tosudite), carbonates (calcite, dolomite, and ankerite), sulphates (anhydrite and barite)

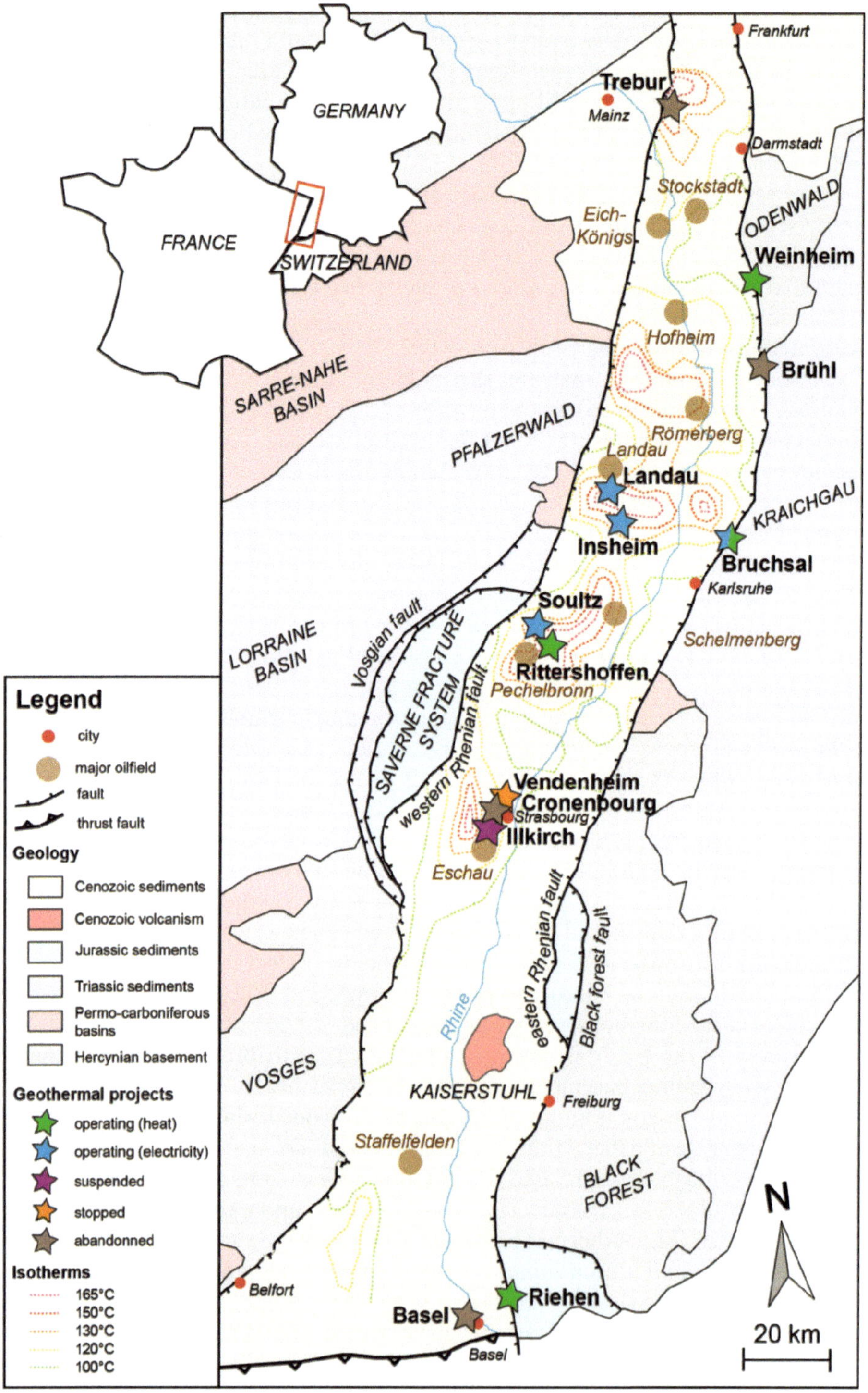

Figure 2.1 Map of the main geothermal sites located in the Upper Rhine Graben

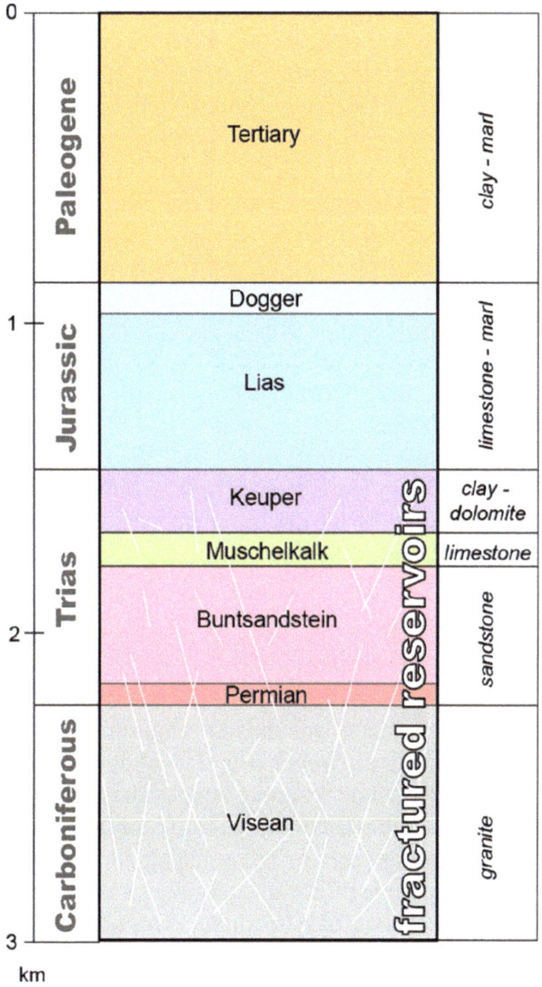

Figure 2.2 Simplified geological section showing the main lithostratigraphic units in the URG

or sulphides (galena, pyrite and chalcopyrite). It is noticeable to observe the absence of kaolinite in hydrothermally altered facies. Based on borehole image logs, temperature profiles, and spinner logs collected in the geothermal wells, a rather limited number of fractures or hydrothermally altered and fractured zones match spatially with discrete outflow anomalies (Evans et al., 2005; Vidal et al., 2015).

Therefore, all the geothermal wells drilled at Soultz crosscut the same crystalline basement characterized by the occurrence of natural brines related to fractured zones where local convection cells take place. Thus, as a lesson learned in this area, the top basement, meaning the first kilometre of crystalline rocks, becomes a real geothermal target and is considered as a geothermal reservoir made of fractured zones surrounded by damage zones rich in secondary clay minerals like illite, a dioctahedral mica rich in potassium (Glaas et al., 2021).

All the industrial geothermal projects located in the URG derived from the Soultz research site (Insheim, Landau, and Rittershoffen) reached their geothermal targets in terms of temperature and permeability. They demonstrated the occurrence of a geothermal resource related to the natural fracture system, mainly located at the interface between the deep Triassic sediments and the top of the crystalline basement. Due to low initial permeability conditions, some geothermal wells were hydraulically and chemically stimulated, whereas other wells were naturally permeable enough to avoid any stimulation. Therefore, the occurrence of this shallower geothermal resource has a serious positive impact on drilling costs. The recently drilled wells in Rittershoffen between 2012 and 2014 confirmed permeability related to natural fractures located within Triassic sedimentary layers (Muschelkalk and Buntsandstein). Observations made on temperature logs in Soultz as well as in Rittershoffen seem to indicate that a no-flow boundary is located at the basis of the Keuper layers (i.e., at the top of the Muschelkalk) and that natural hydraulic flows below these layers are dominated by complex convective circulations.

Therefore, EGS operating plants in the URG are mainly associated with convective, naturally fractured reservoirs. It means that a network of fractures and faults poorly connected or poorly known from the surface could be exploited by deep drilling with borehole trajectories optimally oriented.

2.3 EGS VERSUS HYDROTHERMAL SYSTEM: NATURALLY FRACTURED RESERVOIRS

The first major discovery at Soultz was the abundant presence of natural fluids at great depth. It corresponds to a salty brine having a total dissolved salts (TDS) higher than 100 g/L (Sanjuan et al., 2016). The resource sought then evolved from hot rock to that of a hot fluid. The second major discovery is that the initial permeability at depth is potentially much greater than expected. Indeed, deep crystalline rocks have shown that they can contain important natural drains, namely a vast network of natural fractures. The remaining difficulty was twofold: to connect this natural permeability to the boreholes and to increase the initial permeability of these deep crystalline rock masses by implementing hydraulic, thermal and/or chemical development techniques.

These characteristics have led to this type of deep geothermal energy being known by the acronym EGS. Indeed, it is necessary to act on the near-well bore environment to improve the hydraulic performance of the well, which is often too low to achieve the economic conditions required during the long-term exploitation of the resource. This near-well permeability increase of the fractured rock mass will improve the connection of the open section of the wells to the natural network of fractures. This shear rearrangement is accompanied by induced microseismic activity and facilitates the passage of fluids, which therefore increases the post-stimulation permeability, even if the zones that have ruptured are not necessarily those that are the most initially permeable (Cornet et al., 1997).

This term EGS does not cover the same notions depending on the authors. Indeed, there is a continuum between an EGS geothermal system where the heat exchanger is created from scratch (total absence of initial permeability) and weakly/mediumly hydrothermal systems where there is a fractured aquifer with natural permeability but susceptible to being improved to enable the achievement of industrial operating criteria. It is for this second type that several sites have been developed in the Rhine Graben (Soultz, Landau, Insheim, and Rittershoffen). The resource and its reservoir exist, but it is sometimes necessary to implement these stimulation techniques to improve the connection between the geothermal well and the underground exchanger made of natural fractures to ensure optimized hydraulic performance. Recent development of projects in Northern Alsace shows that the resource could be qualified for EGS because thermo-hydro and chemical stimulation operations were done in the injection well at Rittershoffen (Baujard et al., 2017). This combination of stimulation techniques allows to improve the initial hydraulic performance by a factor of 5 for reaching 2.5 L/s/bar. Only the thermal and hydraulic stimulations generated some induced seismicity at depth, but with seismic events having very low magnitudes, far below what can be felt (Maurer et al., 2020). During the chemical stimulation, a coiled tubing system was used for injecting chemicals able to dissolve mainly secondary carbonates, plugging the natural fractures. Three zones were stimulated chemically in the injection well: two in the deep granite and one in the overlying Triassic sandstone. This chemical stimulation that used packers did not generate any induced seismicity, maybe because secondary mineral dissolution is not creating a hydromechanical effect. Moreover, we used biodegradable chemicals like glutamate acid for such focused stimulation in order to respect the environment even at great depth. The second well at Rittershoffen is purely hydrothermal because a stimulation operation was not necessary. It shows immediately during preliminary well testing that the well has very good hydraulic performance. Thus, the natural permeability shows an industrial flow rate compatible with an economic level and ranges at 4 L/s/bar for this second well.

2.4 INDUCED SEISMICITY IN THE URG BASED ON OPERATIONAL SITES: STATE OF THE ART

Due to its geology and history, the URG today concentrates, at European level, a large number of EGS. These sites have the particularity of having to be stimulated at depths, beyond 2 km. These operations allow the development of the connections between the well and the naturally fractured reservoir to improve the permeability of the underground reservoir and thus facilitate the circulation of the geothermal brine and heat transfer. As a result, seismic monitoring is set up to observe the behaviour of the reservoir. Initially, this monitoring focused on the characterization of low magnitude induced seismicity to image the fracture network and faults in the reservoir activated by the stimulations. With the evolution of the projects and the analysis of the data acquired, the induced seismicity clearly appeared as a dynamic phenomenon that did not only react synchronously to the injections but could still manifest itself after the latter and be felt by the population. As a result, monitoring has adapted to meet the need to minimize seismic hazards throughout the life of an EGS site (Gaucher and Maurer, 2021). On the French side, regional regulations have been put in place and have imposed monitoring, distinguishing the development phase, which includes stimulations, from the operating phase. Today, seismic monitoring includes an alert system based on thresholds of magnitude or peak ground velocity, which implies, if necessary, modifying the operations parameters on site and goes as far as a progressive shut-in of the operations in progress.

EGS technology consists of increasing the low natural hydraulic performance of deep geothermal reservoirs by thermal, chemical, and/or hydraulic stimulations. These stimulations aim to improve the connection of the wells with the nearby formation and to increase the permeability in the reservoir, allowing the geothermal brine to be produced or reinjected at economically viable flow rates. During the drilling or in the course of these stimulations, seismicity is often induced (Maurer et al., 2020), which, on the one hand, provides insight into the reservoir properties but, on the other hand, must be mitigated to prevent harmful effects on the population and goods and on the project's sustainability and acceptability. Seismicity can also be temporarily or continuously induced during the exploitation itself, i.e., during the interwell circulation of the geothermal fluid (Maurer et al., 2021). Since 2016, the geothermal plants of SsF and Rittershoffen have been almost continuously operating, with an availability rate higher than 90%. The induced microseismic activity has been carefully monitored in real time since the beginning of the production. Since then, a couple of thousands of induced low magnitude earthquakes have been detected, all located in the vicinity of the injection wells.

For the Rittershoffen plant, since the plant's commissioning, more than 3,500 microseismic events have been automatically detected and located in the direct vicinity of the project by the local seismic network (Figure 2.3).

All events have been manually relocated. Most events are located within a radius of 1 km around the open-hole section of the injection well GRT-1. The maximum recorded local magnitude Mlv was 1.8 for an event that occurred on September 2021. The maximum PGV is less than 1 mm/s associated with that event. No induced seismic event has been felt by the local population around the Rittershoffen plant since operation started. In SsF, induced seismicity has been observed in injection wells GPK-3 and GPK-4. More than 350 events have been relocated in the vicinity of the project since the commissioning of the plant in July 2016. The maximum local magnitude (Mlv) was 1.7. The maximum PGV associated with that event was 0.173 mm/s. Like the Rittershoffen plant, induced seismicity has not been felt by the local population around SsF since operation started in 2016. However, even if the induced seismicity observed in Rittershoffen and in SsF was not felt by the population, it highlights the need to perform continuous seismic monitoring (Figure 2.4), even during the exploitation phase of geothermal plants, which is now explicitly required by the French mining authorities for getting the exploitation licence (concession).

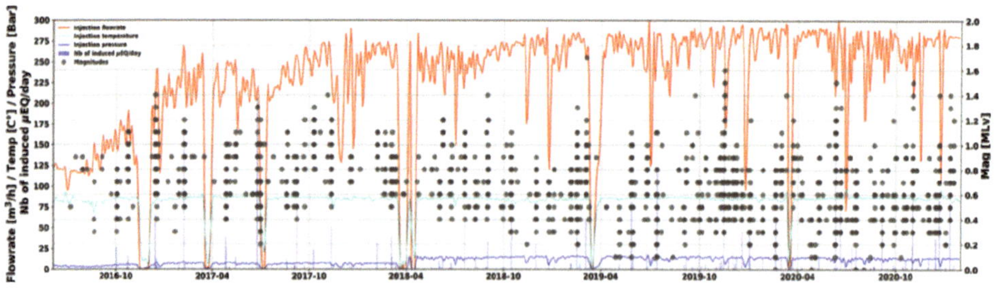

Figure 2.3 Induced seismicity activity (black dot) during geothermal exploitation at Rittershoffen geothermal site from 2016 to 2020

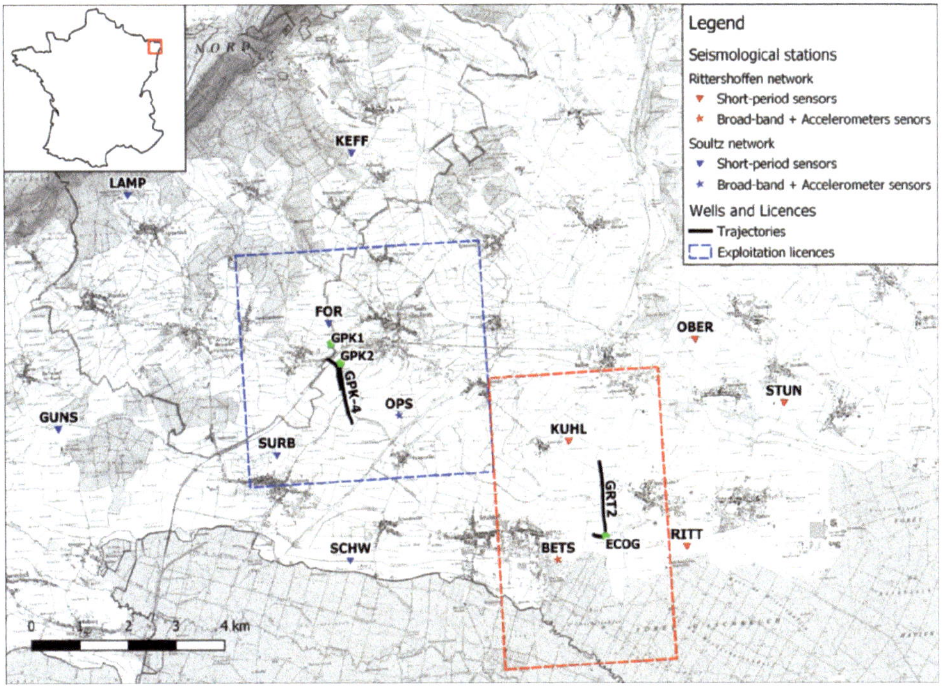

Figure 2.4 Seismological networks of Soultz-sous-Forêts (in blue) and of Rittershoffen (in red) geothermal sites plus the geodetic network (in green). The horizontal traces of the well trajectories are also displayed. Dashed rectangle shows Soultz-sous-Forêts (in blue) and Rittershoffen (in red) concessional areas.

2.5 SUSTAINABLE EXPLOITATION OF EGS PLANTS IN THE URG

Geothermal exploitation of such sites in the URG is a reality. For example, the Soultz and Rittershoffen sites have been operating since mid-2016 without any problems. Availability is over 90%, meaning that there are less than 20 days of maintenance per year since the plants commissioning in 2016.

Currently, the Soultz geothermal site exploits around 30 L/s of geothermal water with a very high salinity of 100 g/L. The brine is produced at 150°C from the production well GPK-2 and conveyed after filtration to three heat exchangers, supplying heat to a 1.7 MW ORC (Organic Rankine Cycle) unit. This electricity production unit is the only heat user of this geothermal plant and uses the ambient air as a cooling medium through an air condenser. The ambient air temperature variations throughout the day and the year impact the geothermal brine reinjection temperature, ranging from 65°C in winter to 80°C in summer. The geothermal brine is reinjected into two different wells, GPK-3 and GPK-4. The geothermal brine is produced at a temperature of 150°C, reaching the wellhead with a nominal flow rate of 30 kg/s provided by a downhole production line shaft pump (LSP). A view of the SsF geothermal power plant is given in Figure 2.5. The wellhead overpressure in the surface infrastructure is controlled by

Figure 2.5 The Soultz-sous-Forêts power plant operating in Northern Alsace, France

the downhole production pump and reaches about 22 bars to keep the dissolved gas in the brine. The geothermal plant has been successfully producing electricity since September 2016 under commercial conditions, with an availability rate of about 90% for the last 5 years.

The geothermal heat plant at Rittershoffen has been developed to supply heat to the industrial processes of a starch plant. The Rittershoffen heat plant has an installed capacity of 27.5 MWth, and it has been providing an average of 22.5 MWth and 180 GWh/year of heat to the heat user since June 2016. The Rittershoffen heat plant reservoir is a Triassic sandstone and the top of a fractured carboniferous granite basement located at 2,500 m depth. The heat plant relies on two wells. Due to the nature of the reservoir, the heat plant was designed with a pressurized geothermal loop. The loop consists of a downhole LSP, which works to pressurize the geothermal brine in the surface equipment to exceed the gas break-out pressure. This pressure allows gas to remain dissolved in the liquid phase and prevents any NCG emissions during operation. The production wellhead temperature reaches 170°C, and the flow rate is regulated at 75–85 kg/s, according to the starch plant's heat demand. The geothermal heat is transferred to a secondary loop containing fresh water using several tubular heat exchangers, and the brine is fully reinjected without additional pumps at 85°C into the injection well. The Rittershoffen plant availability has been 92% higher for the last five years. A typical EGS plant is made of two geothermal wells able to produce heat at high temperatures like Rittershoffen, electricity like in Soultz, and in the future, lithium extraction from the geothermal brine (Figure 2.6).

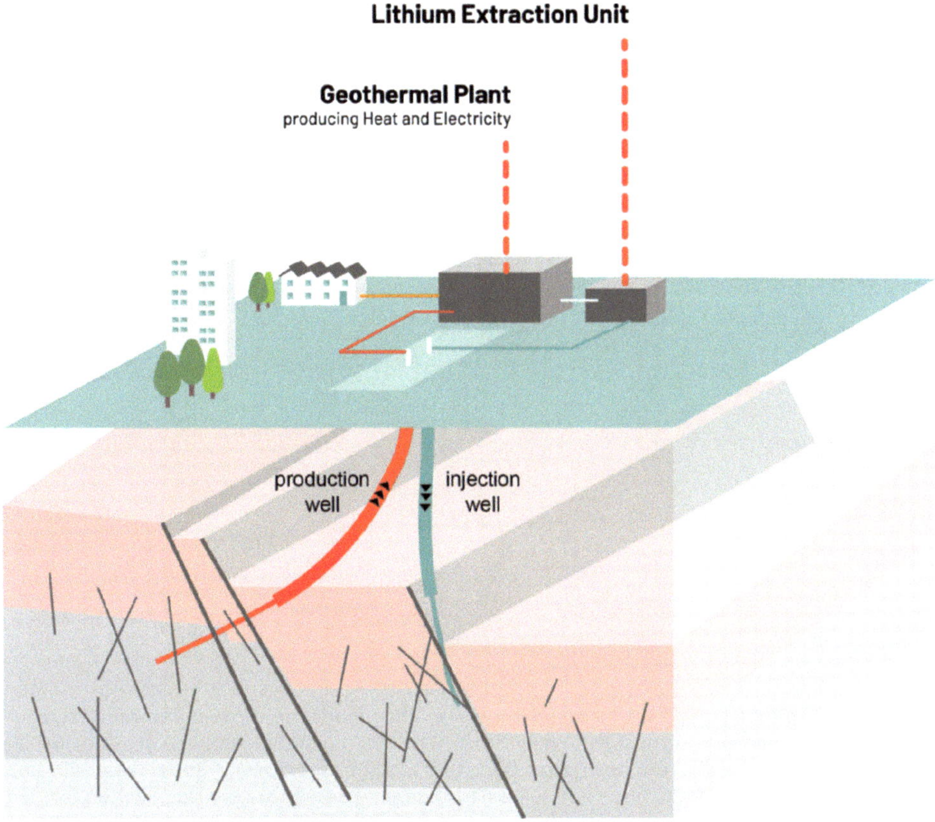

Figure 2.6 Conceptual design of a EGS plant in the URG supplying heat, producing electricity and extraction lithium from its geothermal brines

2.6 EGS AND LIFE CYCLE ASSESSMENT

Geothermal energy is a promising renewable energy source and could play an important role in the decarbonization of the energy sector. To make sure that geothermal energy does not imply additional environmental impacts, it is essential to determine all the potential environmental impacts of such a technology while considering its entire life cycle. Life cycle assessment (LCA) is a standardized tool used to quantify various environmental impacts of a technology or product throughout its entire life cycle (ISO 14040, 2006).

Lacirignola et al. (2013) conducted the first LCA on the EGS plant based on the SsF case study and concluded that the impact on climate change is about 36.7 gCO_2-eq/kWh of electricity. Pratiwi et al. (2018) further assessed the impact on climate change of heat production from the EGS plant based on the case study of the Rittershoffen heat plant. Pratiwi's study was then completed by Douziech et al. (2020), applying LCA guidelines developed by Parisi et al. (2020) to harmonize LCA practices in the

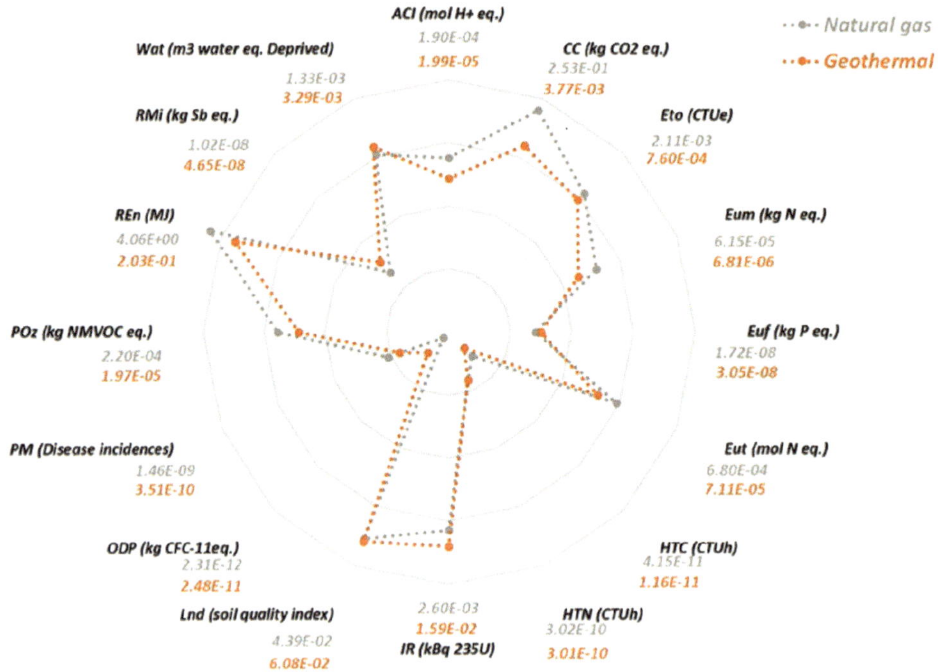

Figure 2.7 Comparison of the production of 1kWhth from the most common heat source in Europe (Natural gas, grey dots) with the investigated heat production from the Rittershoffen heat plant (Geothermal, orange dots), (Douziech et al., 2021). ACI: Acidification, CC: Climate change, Eto: Ecotoxicity, freshwater, Eum: Eutrophication marine, EUf: Eutrophication freshwater, EUt: Eutrophication terrestrial, HTC: Human toxicity cancer, HTN: Human toxicity non-cancer, IR: Ionising radiation human health, Lnd: Land use, ODP: Ozone depletion, PM: Particulate Matter, POz: Photochemical ozone formation - human health, REn: Resource use, fossils, RMi: Resource use minerals and metals, Wat: Water use

geothermal sector. Douziech et al. (2021) reported 3.8 gCO_2-eq/kWhth for the Rittershoffen heat plant, about 67 times lower than that of heat from biomass or natural gas. By replacing natural gas in its energy mix, the starch plant reduces its greenhouse gas emissions by over 40,000 tons of CO_2-eq per year. Figure 2.7 also shows that geothermal heat production performed better on 9 out of the 16 impact categories selected, showing that EGS plants can provide sustainably high levels of heat with lower or comparable environmental impacts when using natural gas.

2.7 HOW TO DECREASE REINJECTION TEMPERATURE FOR PRODUCING MORE ENERGY FROM EGS?

Most of the geothermal plants in the URG are reinjecting natural brines into the reservoir at a temperature of about 80°C. It means that there is a large amount of energy still available that is not being used. In the frame of a R&D research project, MEET, a small-scale unit ORC operating below 80°C was used in order to exploit the thermal spectrum towards lower temperatures (Dalmais et al., 2021). A feasibility study and on-site tests on the SsF power plant were conducted to decrease the temperature at reinjection from 80°C to 40°C in order to assess energy valorization by minimizing thermal losses. Different tests were conducted on the SsF power plant to achieve this demonstration. The first test of temperature decrease has been attempted on 10% of the flow rate during three months between February and April 2019 (Seibel et al., 2021). In parallel, environmental studies like induced microseismicity monitoring during cold reinjection were conducted as well as reservoir modelling, which must ensure that there will be no significant thermal breakthrough or flow rate decrease during the lifetime of the exploitation (Baujard et al., 2021).

Without considering the effect of further decreasing the brine injection temperature on the reservoir and the propagation of the cold plum towards the production well, lowering the current limit for the low temperature of brine injection to about 50°C seems feasible. Since the variations in the geochemistry of the brine produced in the URG do not greatly differ from one site to another, it would seem some specific materials like Super Duplex can be recommended for heat exchangers or other equipment used in all the existing and future geothermal plants operating in this area. Table 2.1 presents the potential thermal valorization that could be achieved by assuming a potential lowering of the brine injection temperature to 50°C in order to keep an operational margin at the point at which the deposition of new amorphous scaling containing silica theoretically occurs.

2.8 NATIVE FLUID GEOCHEMISTRY

In the URG, all the geothermal sites are characterized by salty brines (Sanjuan et al., 2016). For example, at the Rittershoffen site, the geothermal brine is a Na-Ca-K-Cl–dominated brine with a TDS content of approximately 107 g/L. Non-Condensable Gas (NCG) content is 0.24% in weight mass, and it is composed mainly of CO_2.

Several studies have focused on the characterization of the geochemical brine from the URG (Sanjuan et al., 2016), providing a large database for major and trace element concentrations. However, those data are mainly measured on the brine collected after the drilling operation and development, while very few characterizations concern the geochemical evolution of the brine during several years of continuous operation (Bosia et al., 2021). The geothermal fluids of Rittershoffen and SsF plants are Na-Ca-Cl brines with important concentrations of K. As already shown by several studies (Sanjuan et al., 2016; Bosia et al., 2021), this composition corresponds to a typical geothermal fluid circulating in the basement reservoir of the URG. The

Table 2.1 Potential thermal valorization in existing and future geothermal plants in the URG

Site	*Production flow (L/s)*	*Production temperature (°C)*	*Injection temperature (°C)*	*Current thermal power used (MW)*	*Valorization potential at 50°C (MW)*
Soultz-sous-Forêts	30	150	65	11.3	1.7
Rittershoffen	75	168	80	25	5.6
Insheim	70	164	65	26	4
Landau	70	155	65	24	4
Bruchsal	24	123	65	5.5	1.4

concentration of those major ions is almost identical between the two geothermal sites of SsF and Rittershoffen, and no significant variations are observed versus time over the exploitation, neither at the production well nor at the injection well. It means that the geothermal reservoir is relatively large because the horizontal distance between those two sites is about 8 km.

The homogeneity of the chemical signature of the brine at the production and injection wells indicates that geothermal fluid circulation in the surface equipment does not significantly affect the composition of the fluids, despite the existence of corrosion, scale formation and degassing processes in the installations (Mouchot et al., 2018). However, the absence of sensitive geochemical modification of the fluid highlights that corrosion and scale processes are today well controlled by the site operator, thus limiting their impacts on the geothermal infrastructure, in terms of costs, efficiency, and operability of the plants.

2.9 THE FUTURE OF GEOTHERMAL ENERGY: ITS METAL-RICH BRINE

The unaltered granite encountered by geothermal wells in Northern Alsace is a porphyritic monzogranite with K-feldspar megacrysts containing primary biotite, hornblende, oligoclase, and quartz. There are accessory minerals like magnetite, titanite, apatite, zircon, and leucoxene (Figure 2.8).

The lithium content in the bulk rock is about 40 ppm in such an unaltered monzogranite facies. Due to the complex geological history from its emplacement at Carboniferous time, the granitic basement rocks underwent several events of hydrothermal alteration: (i) a propylitic alteration that developed pervasively in the granitic pluton during the cooling subsequent to its emplacement and (ii) an argillic alteration that developed in response to intense fluid/rock interactions triggered by fluid circulation through the fracture network. The main mineralogical signature of the argillic alteration is characterized by the occurrence of clay minerals like illite, a dioctahedral clay bearing potassium. Locally, tosudite, a clay mineral corresponding to an ordered dioctahedral interstratified chlorite/smectite mineral bearing lithium, has been observed at Soultz in the hydrothermally altered and fractured granite basement. This secondary clay bearing lithium, which was spatially associated with the occurrence

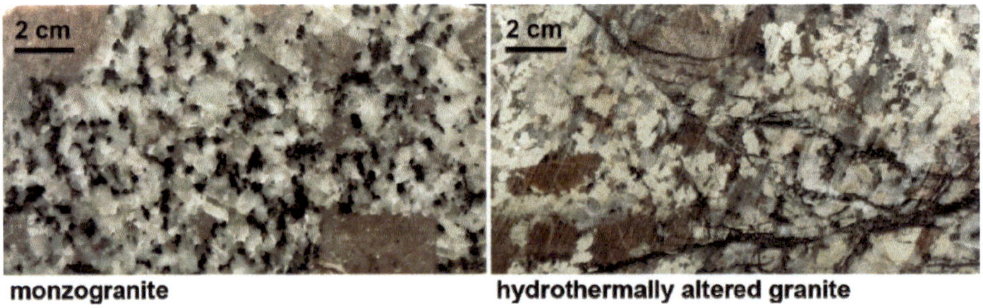

Figure 2.8 The Soultz-sous-Forêts monzogranite. The unaltered massive monzogranite showing the coarser K-felspar megacrysts (left) and a hydrothermally altered and fractures monzogranite (right)

of organic matter, has been observed in an altered granite sample showing 1,447 ppm of lithium (Ledésert et al., 1999). Other secondary minerals could precipitate as fracture filling and correspond mainly to geodic quartz, carbonates, or sulphates and sulphides. There are very few chemical analyses of the bulk rocks in altered granite for lithium. Generally, lithium content is highly variable and ranges from 22 to 244 ppm, with a maximum of 1,447 ppm (Ledésert et al., 1999).

Lithium and other elementary chemical measurements in a hydrothermally altered granite facies affected by argillic alteration are presented in Figure 2.9. It corresponds to the mapping of chemical elements measured with the Laser-Induced Breakdown Spectroscopy (LIBS) method. For instance, within this altered facies, primary biotite shapes are preserved, as evidenced by noticeable concentrations in both iron and titanium. However, these relics of primary biotites are depleted of lithium. Altered plagioclases are marked by the presence of calcium, but they also contain noticeable concentrations of lithium. Barium is mainly concentrated in K-Felspar megacrysts.

Therefore, this depletion of lithium within altered biotites, which normally bear lithium, could be interpreted as a chemical migration taking place during the hydrothermal alteration process related to fluid circulation in fractures. Thus, lithium is concentrated in the altered plagioclases, which are mainly composed of secondary illites.

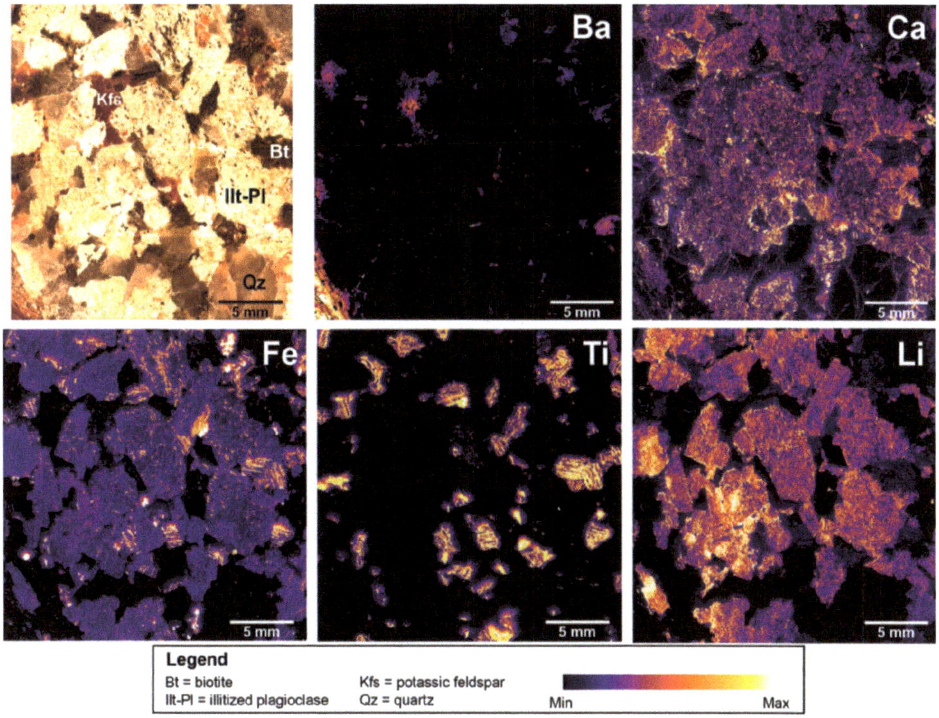

Figure 2.9 Elementary chemical maps for Ba, Ca, Fe, Ti and Li of a hydrothermally altered granite sample taken in a geothermal well in Northern Alsace, France.

Among the other elements composing the chemical signature of the Rittershoffen and SsF fluids, some constituents reach concentrations of hundreds of ppm, in particular dissolved lithium. It has been known for decades that the brines in the URG are strongly enriched in metals like lithium, with concentrations around 168 and 181 ppm for SsF and Rittershoffen, respectively. Moreover, it has been shown that the lithium concentration is very high and stable over time during exploitation. Thus, recent research conducted on lithium extraction during exploitation at SsF has proved that about 90% of the lithium could be extracted from the brine. It means that for a flow rate of about 85 L/s and an availability of 95%, the lithium quantity potentially extractable per year as lithium carbonate equivalent could be about 1,500 tons at Rittershoffen. Thus, it represents a real challenge for electric vehicles with batteries using lithium-ion technology in the future.

The similar composition of all the geothermal fluids from the URG is an indicator of their common geological origin. Until now, the geothermal brine exploited for heat and power production in the region was issued from a geothermal reservoir located at the granitic basement interface, whose permeability is provided by the fracture network and by the secondary porosity of the granitic matrix (Genter et al., 2010; Glaas et al., 2021). With Li concentrations higher than 150 ppm, recent studies show that the URG could be one of the most promising zones in Europe for exploiting lithium dissolved in the geothermal brines (Bosia et al., 2021; Sanjuan et al., 2022). Based on fluid geochemistry and isotopic data, the main source of lithium is assumed to be the 400-m-thick micaceous continental sandstone of Triassic Buntsandstein, even though a contribution from the granite basement is not excluded (Sanjuan et al., 2022). By taking into account many assumptions, namely the rock porosity of the Triassic reservoirs, a Li resource assessment done on a block of 30 km × 30 km centred around SsF estimated a quantity of lithium ranging from 300,000 to 2,200,000 tons of lithium metal (Pauwels et al., 1991). Those estimations did not take into account the potential contribution of the deep, fractured granite basement.

2.10 CONCLUSION

EGS development in Europe and mainly in the URG, lessons learned and characterization of the underground conducted to consider naturally fractured zones or faults as the main geothermal reservoirs. Natural fluids with similar geochemical composition might be interpreted as the occurrences of large-size geothermal reservoirs, which consist of fluid circulation through the fractured granitic basement and its Triassic overlying sedimentary cover. However, achieving an economic flow rate could necessitate thermo-hydro or chemical stimulation techniques for enhancing the hydraulic performance of the geothermal wells. Moreover, geothermal exploitation over several years of URG EGS plants with low environmental impact has fully demonstrated that geothermal energy could produce renewable electricity, heat at high temperatures for industry, and tomorrow neutral carbon lithium for batteries. The main challenge for EGS is mainly the potential occurrence of induced seismicity related to stimulation or exploitation, which could generate severe societal concerns. The absence of felt-induced seismicity related to the geothermal exploitation of the SsF and Rittershoffen plants since 2016 is clear proof that there is a route for producing low carbon energy for the local population in Europe.

ACKNOWLEDGEMENTS

The authors would like to thank the SsF site owner and the Rittershoffen site owner, respectively, GEIE Exploitation Minière de la Chaleur and ECOGI. This research has received funding from the European Union's Horizon 2020 research and innovation programme under grant agreement no. 792037 (MEET project). The authors would also like to acknowledge the French research ThermaLi project, cofunded by the GIS Geodenergies and ANR (Agence Nationale de la Recherche). The authors would like to thank Prof Ingrid Stober for her improvement of the manuscript.

REFERENCES

Baujard C., Genter A., Dalmais E., Maurer V., Hehn R., Rosillette R., Vidal J., Schmittbuhl J., (2017). Hydrothermal characterization of wells GRT-1 and GRT-2 in Rittershoffen, France: Implications on the understanding of natural flow systems in the Rhine graben, *Geothermics* 25, 255–268.

Baujard C., Rollin P., Dalmais E., Hehn R., Genter A., (2021). Soultz-sous-Forêts geothermal reservoir: Structural model update and thermo-hydraulic numerical simulations based on 3 years operation data, *Geosciences* 11, 502. https://doi.org/10.3390/geosciences11120502.

Bosia C., Mouchot J., Ravier G., Seibt A., Jähnichen S., Degering D., Scheiber J., Dalmais E., Baujard C., Genter A., (2021). Evolution of brine geochemical composition during operation of EGS geothermal plants (Alsace, France), *Stanford Geothermal Workshop 2021*, Stanford, CA, February 16–18, 2021.

Cornet F.H., Helm J., Poitrenaud H., Etchecopar A., (1997). Seismic and aseismic slips induced by large-scale fluid injections, *Pure Applied Geophysics* 150, 563–583.

Dalmais E., Genter A., Trullenque G., Leoutre E., Leiss B., Bär K., Mintsa A.-C., Ólafsson D.I., Rajsl I., Wagner B., (2021). MEET Project: Toward large scale deployment of deep geothermal energy in Europe, *World Geothermal Congress 2020*, WGC 2020+1, Reykjavik, Iceland, October 24–27, 2021.

Dempsey D., Kelkar S., Davatzes N.C., Hickman S., Moos D., Zemach, E., (2014). Evaluating the roles of thermoelastic and poroelastic stress changes in the desert peak EGS stimulation, in *Proceedings of Thirty-Ninth Workshop on Geothermal University*, Stanford University, California.

Douziech M., Ravier G., Jolivet R., Perez Lopez P., Blanc I., (2020). How far can life cycle assessment be simplified? A protocol to generate simple and accurate models for the assessment of energy systems and its application to heat production from enhanced geothermal systems, *Environmental Science & Technology* 55 (11), 7571–758. https://pubs.acs.org/doi/full/10.1021/acs.est.0c06751.

Douziech M., Tosti L., Ferrara N., Parisi M.L., Pérez-López P., Ravier G., (2021). Applying harmonised geothermal life cycle assessment guidelines to the Rittershoffen Geothermal Heat Plant, *Energies* 14, 3820. https://doi.org/10.3390/en14133820.

Evans K.F., Genter A., Sausse J., (2005). Permeability creation and damage due to massive fluid injections into granite at 3.5 km at Soultz: part 1 - Borehole observations, *Journal of Geophysical Research* 110, B04203.

Gaucher E., Maurer V., (2021). Évolution du suivi sismique des sites géothermiques en exploitation dans le Fossé rhénan supérieur, *Géologues Revue Géologues* 210, 94–102.

Genter A., Evans K., Cuenot N., Fritsch D., Sanjuan B., (2010). Contribution of the exploration of deep crystalline fractured reservoir of Soultz to the knowledge of enhanced geothermal systems (EGS), *Comptes Rendus Geoscience* 342, 502–516. https://doi.org/10.1016/j.crte.2010.01.006.

Gérard A., Kappelmeyer O., (1987). The Soultz-sous-Forêts project, *Geothermics* 16, 393–399. https://doi.org/10.1016/0375-6505(87)90018-6.

Glaas C., Patrier P., Vidal J., Beaufort D., Genter A., (2021). Clay mineralogy: A signature of granitic geothermal reservoirs of central Upper Rhine Graben, *Minerals* 11, 479. https://doi.org/10.3390/min11050479.

Kohl Th., Bächler D., Rybach L., (2000). Steps towards a comprehensive thermo-hydraulic analysis of the HDR test site Soultz-sous-Forêts, *WGC2000, World Geothermal Congress*, Kyushu-Tohoku, Japan, May 28–June 10, 2000, pp. 2671–2676.

Lacirignola M., Blanc I., (2013). Environmental analysis of practical design options for enhanced geothermal systems (EGS) through life-cycle assessment, *Renewable Energy* 50, 901–914.

Ledésert B., Berger G., Meunier A., Genter A., Bouchet A., (1999). Diagenetic-type reactions related to hydrothermal alteration in the Soultz-sous-Forêts granite, France, *European Journal of Minerology* 11, 731–741.

Maurer V., Cuenot N., Richard A., Hehn R., (2021). Geophysical monitoring of EGS fields in the Upper Rhine Graben, *Proceedings of World Geothermal Congress 2021*, WGC2020+1, , Online & On-site in Reykjavik, Iceland, March–October 2021.

Maurer V., Gaucher E., Grunberg M., Koepke R., Pestourie R., Cuenot N., (2020). Seismicity induced during the development of the Rittershoffen geothermal field, France, *Geothermal Energy* 8, 5. https://doi.org/10.1186/s40517-020-0155-2.

Mouchot J., Genter A., Cuenot N., Scheiber J., Seibel O., Bosia C., Ravier G., (2018). First year of operation from EGS geothermal plants in Alsace, France: Scaling issues, *Proceedings, 43rd Workshop on Geothermal Reservoir Engineering*, Stanford University, Stanford, CA.

Nami P., Schellschmidt R., Schindler M., Tischner T., (2008). Chemical Stimulation operations for reservoir development of the deep crystalline HDR/EGS system at Soultz-sous-Forêts (France), in *Proceedings of Thirty-Second Workshop on Geothermal Reservoir Engineering*, Stanford University, Stanford, CA, p. 5.

Parisi M.L., Douziech M., Tosti L., Pérez-López P., Mendecka B., Ulgiati S., Fiaschi D., Manfrida G., Blanc I., (2020). Definition of LCA Guidelines in the geothermal sector to enhance result comparability. *Energies* 13, 3534.

Pauwels H., Lambert M., Genter A., (1991). Valorisation des fluides géothermaux contenant du lithium en vue d'une production industrielle (in French), Orléans, Open File BRGM Report, R 33547, 173 p + annexes.

Portier S., Vuataz F.-D., Nami P., Sanjuan B., Gérard A., (2009). Chemical stimulation techniques for geothermal wells: experiments on the three-well EGS system at Soultz-sous-Forêts, France, *Geothermics* 38, 349–359.

Pratiwi A., Ravier G., Genter A., (2018). Life-cycle climate-change impact assessment of enhanced geothermal system plants in the Upper Rhine Valley, *Geothermics* 75, 26–39. https://doi.org/10.1016/j.geothermics.2018.03.012.

Sanjuan B., Gourcerol B., Millot R., Rettenmaier D., Jeandel E., Rombaut A., (2022). Lithium-rich geothermal brines in Europe: An up-date about geochemical characteristics and implications for potential Li resources, *Geothermics* 101, 102385. https://doi.org/10.1016/j.geothermics.2022.102385.

Sanjuan B., Millot R., Innocent Ch., Dezayes Ch., Scheiber J., Brach M., (2016). Major geochemical characteristics of geothermal brines from the Upper Rhine Graben granitic basement with constraints on temperature and circulation, *Chemical Geology* 428, 27–47. https://doi.org/10.1016/j.chemgeo.2016.02.021.

Schill E., Genter A., Cuenot N., Kohl Th., (2017). History performance history at the Soultz EGS reservoirs from stimulation and long-term circulation tests, *Geothermics* 70, 110–124.

Schindler M., Baumgartner J., Gandy T., Hauffe P., Hettkamp T., Menzel H., Penzkofer P., Teza D., Tischner T., Wahll G., (2010). Successful hydraulic stimulation techniques for electric power production in the Upper Rhine Graben, Central Europe, in *Proceedings of World Geothermal Congress 2010*, WGC2010, Bali, Indonesia.

Schmittbuhl J., Lambotte S., Lengliné O., Grunberg M., Jund H., Vergne J., Cornet F.H., Doubre C., Masson F., (2021). Induced and triggered seismicity below the city of Strasbourg, France from November 2019 to January 2021, *Comptes Rendus. Géoscience* 353, 561–584. https://doi.org/10.5802/crgeos.71.

Seibel O., Mouchot J., Ravier G., Ledesert B., Sengelen X., Hebert R., Ragna Ragnarsdottir K., Ingi Olafsson D., Osk Haralddottir H., (2021). Optimised valorisation of the geothermal resources for EGS plants in the Upper Rhine Graben, *World Geothermal Congress 2020*, WGC 2020+1, Reykjavik, Iceland, October 24–27, 2021.

Vidal J., Genter A., (2018). Overview of naturally permeable fractured reservoirs in the Upper Rhine Graben: Insights from geothermal wells, *Geothermics* 74, 57–73. https://doi.org/10.1016/j.geothermics.2018.02.003.

Vidal J., Genter A., Schmittbuhl J., (2015). How permeable fractures in the Triassic sediments of Northern Alsace characterize the top of hydrothermal convective cells? Evidences from Soultz geothermal boreholes (France), *Geothermal Energy. Special Issue: Characterization of Deep Geothermal Systems* 3, 8. https://doi.org/10.1186/s40517-015-0026-4.

Vidal J., Genter A., Schmittbuhl J., (2016). Pre- and post-stimulations of the geothermal well GRT-1 (Rittershoffen, France): Insights from acoustic image logs on hard fractured rock investigations, *Geophysical Journal International* 206, 845–860.

Vuataz F-D., Brach M., Criaud A., Fouillac Ch., (1990). Geochemical monitoring of drilling fluids: A powerful tool to forecast and detect formation waters, *SPE Formation Evaluation* 5 (2), 177–184.

Chapter 3

High radiogenic granites of western Anatolia for EGS

A review

Dornadula Chandrasekharam, Alper Baba, and Tolga Ayzit
Izmir Institute of Technology

CONTENTS

> Focus on electricity decarbonisation. Strong policies supporting low carbon electricity could more than halve electricity emissions in 2030. This would save 9.5 Gt in 2030 – an amount larger than China's total 2012 energy emissions.
>
> International Energy Agency 2014

3.1 INTRODUCTION

Energy is a key factor in the socio-economic development of any country. Energy secures food and water independence for a country. Countries' can plan their sustainable development strategy and maintain a high GDP if and only if a sound food-water and energy security scenario is established. This is termed the food-energy-water nexus. Many countries depend on either imported or indigenous fossil fuels to meet their primary energy needs. As the demand for energy rises due to population growth, the use of fossil fuels also increases. Fossil fuels being the main source of CO_2 emissions, the quantity of CO_2 emissions also increases with an increase in the usage of fossil fuels.

DOI: 10.1201/9781003271475-3

During the last year, globally, energy demand has increased by over 4.6% (IEA, 2021a). The use of fossil fuels has an adverse effect on the environment and does not support the sustainable development of any country. Over the years, the GHG emissions emitted by fossil fuels have had an adverse effect on the climate, causing irreparable damage to the environment and population. Recently, the countries have experienced forest fires, flash floods and extreme temperatures. These have been attributed to an increase in global temperatures and climate change caused by an increase in CO_2 emissions, especially from the use of fossil fuels for energy supply and transportation (UNFCCC, 2015; IEA, 2021). The Conference of Parties, a group of nearly 195 countries, in their (UNFCCC, 2015) meeting held in Paris in 2015, reached a historic agreement and accepted a resolution to adopt a strategy to reduce CO_2 emissions by evolving alternate carbon-free or low-carbon-emitting energy sources to replace a significant part of fossil fuels as a primary energy source. Turkey has accepted this agreement and implemented actions to reduce the use of fossil fuels. Currently, the carbon-free or low-carbon emitting renewable energy sources are wind, nuclear and geothermal. These sources are currently deployed in certain countries, but their contribution to the primary energy supply is low. Although solar PV (photovoltaic) is also considered a carbon-free energy source, it cannot be considered a low-carbon source due to its high emissions during its life cycle (Chandrasekharam and Ranjith, 2020). Turkey's primary energy source is fossil fuels, accounting for 56% of its contribution to the total energy supply (Table 3.1). Currently, Turkey's emissions are crossing 400 Mt (Figure 3.1), and this will keep increasing with the demand for power due to the increase in population (Figure 3.1). The transport sector is the second largest emitter of CO_2 (Figure 3.2). Although Turkey has tremendous hydrothermal and enhanced geothermal system (EGS) potential, its contribution to the primary energy supply is only small.

This chapter focuses on the current renewable sources that are contributing to the primary energy sector, their current generation and future projections. It is focused mainly on the geothermal sources (hydrothermal and EGS), as this source, compared to other sources like wind and hydro, has a low-carbon footprint, is environmentally friendly, and can be developed anywhere on earth. We have not considered solar PV here, as this source, during its life cycle, emits a considerable amount of CO_2 (Chandrasekharam and Ranjith, 2020). EGS being an emerging technology that harnesses heat from high radiogenic granites, and Turkey being one of the countries with huge volumes of high radiogenic granites distributed along the

Table 3.1 Present and projected energy sources of Turkey

Year	*2020*	*2030*
Fossil fuels (Mtoe)	33	36
Hydro	10	10
Geothermal	1.7	3.6
Biomass	5	5
Solar and wind	2	4
Total production	51.7	58.6

Source: Adapted from Keles and Bilgen, 2012; Chandrasekharam and Baba, 2022.

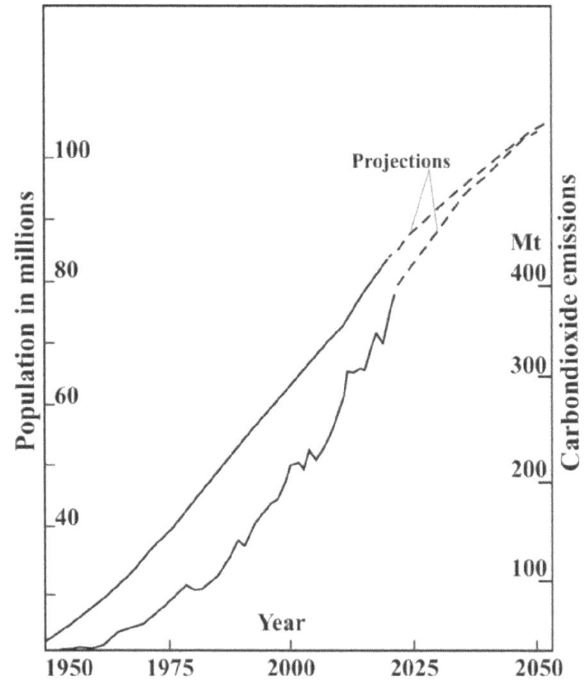

Figure 3.1 Impact of population growth on CO_2 emissions. Adapted from Chandrasekharam and Baba, 2022.

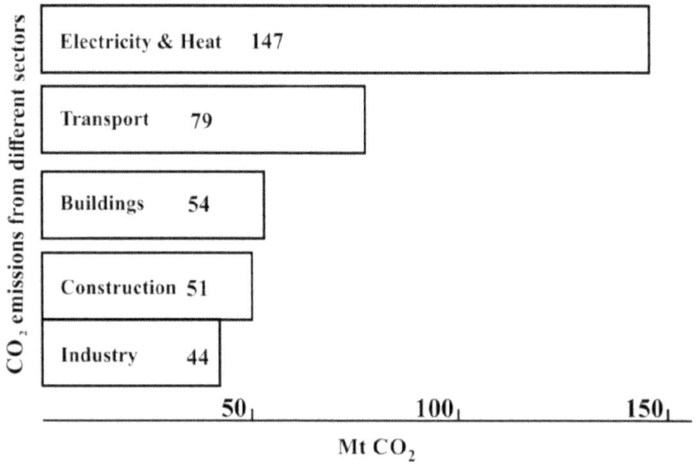

Figure 3.2 CO_2 emissions from different sectors. Adapted from Chandrasekharam and Baba, 2022.

western Anatolian region, our main emphasis in this chapter centres around such granites and heat extraction from high radiogenic granites. The information provided in this chapter is advantageous to the EGS technological developments that are taking place and are highlighted in this book in other chapters.

3.2 METHOD OF APPROACH

The chapter gives brief information related to the current status of geothermal energy utilization in Turkey from its hydrothermal resources spread along the western Anatolian region, followed by a review related to EGS. For EGS projects, the most important components of studies include (i) a sound surface geological information, (ii) regional and local surface structural details like faults and joints, (iii) regional stress regime of the proposed EGS site, (iv) subsurface geological and tectonic fabric based on geophysical investigation, (v) petrofabric and microfabric details of the rocks under consideration for exploration, (vi) geochemistry of the rock (s) that are considered as potential candidate for EGS, (vii) heat generation (radioactive heat production RHP) of the EGS candidate(s), (viii) heat flow values calculated from RHP as well as those measured in the field, (ix) geophysical survey (deep resistivity, magneo-telluric, gravity, Curie point depth survey) and (x) bore hole information if any available. It is considered intrusive granite of Precambrian, Mesozoic, Paleozoic and Cenozoic ages that is outcropping over the entire western Anatolian region, extending up to 36°E latitude (Figure 3.3), with a surface area of 6,910 km^2.

3.3 CURRENT STATUS OF HYDROTHERMAL RESOURCES OF TURKEY

The entire tectonic fabric of Turkey has developed due to the active Himalayan-Alpine orogeny (Bozkurt and Oberhansli, 2001; Baba and Chandrasekharam, 2022; Chandrasekharam and Baba, 2022) that has resulted in a regime of crustal extension and shortening, high geothermal gradient and heat flow values and the formation of horst and graben structures, especially along western Anatolia. These activities have given rise to over 346 geothermal fields in 63 geothermal provinces (Figure 3.4). The temperature of the thermal waters at the surface varies between 20°C and 93°C. Hydrothermal exploration started in 1984, the year when the country established its first geothermal power plant with an installed capacity of 15 MWe (Baba et al., 2021). By the year 2022, Turkey had established a firm ground in geothermal exploration with an installed capacity of (Figure 3.5) 1,663 MWe and had become the fourth top country in the world in geothermal power generation (Hutter, 2020; Baba et al., 2021).

The geothermal wells drilled during these years (1984–Present) provided vital information related to the subsurface temperatures in the western Anatolian region. These wells were drilled both for power projects as well as for direct applications (space heating/cooling, greenhouse cultivation). The bottom hole temperature of the bore-wells varies from 52°C (500 m deep) to 341°C (3950 m deep) (Baba et al., 2021, Baba and Chandrasekharam, 2022; Şener et.al., 2022). While Turkey has 60,000 MWt of hydrothermal potential, only 4,000 MWe are available for power generation (Baba

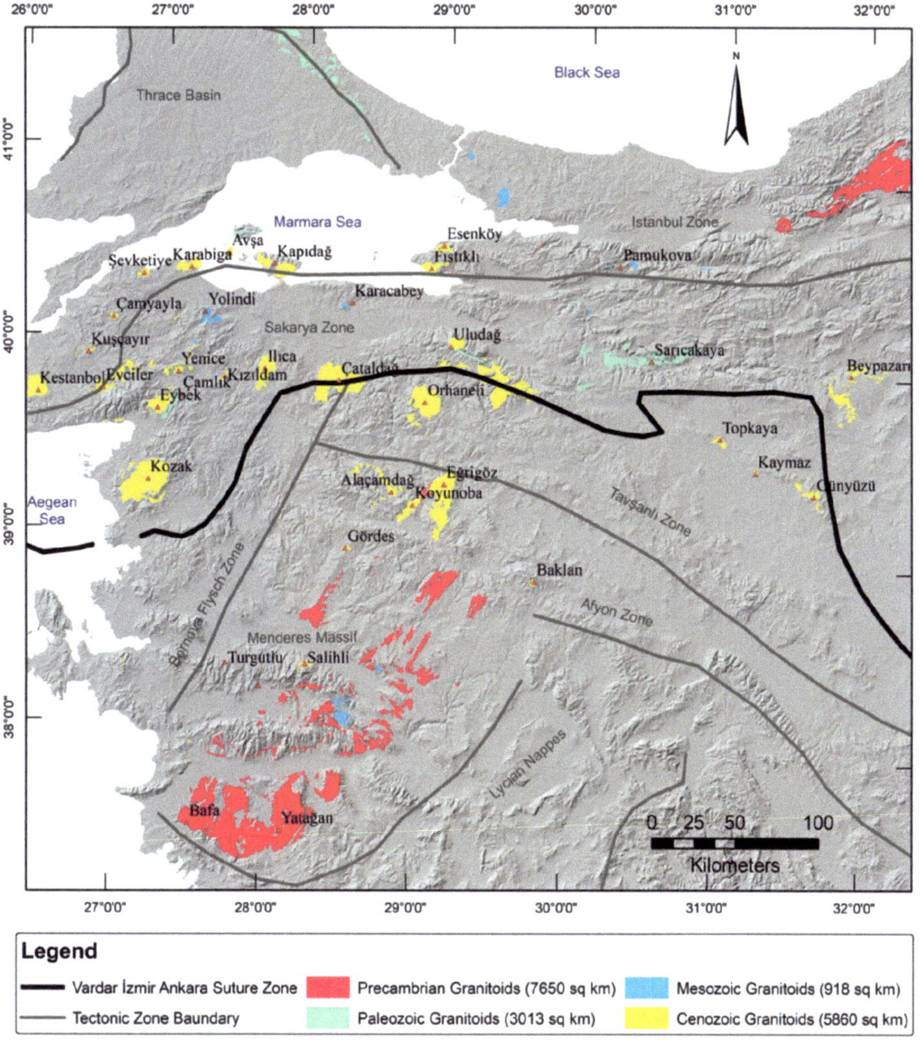

Figure 3.3 Granite intrusive in western Anatolia (for source, see Table 3.2).

et al., 2021; Baba and Chandrasekharam, 2022). While the high-temperature geothermal fields are being utilized for power generation, the medium temperature resources and spent waters from power plants are being utilized for a variety of direct applications like space heating and cooling, greenhouse cultivation and small-scale industries like the cigarette paper manufacturing industry. Thus, 42% of the resources are utilized for space heating and cooling individual residential homes and district heating systems, including spas and hotels, and 8% are utilized for greenhouse cultivation (Figure 3.6). Turkey ranks second in the world in direct application of geothermal resources (Şener et al., 2022) with an installed capacity of 5,113 MWt. Nearly 150,000 homes with an individual area of $100\,m^2$ are currently geothermally heated (Baba

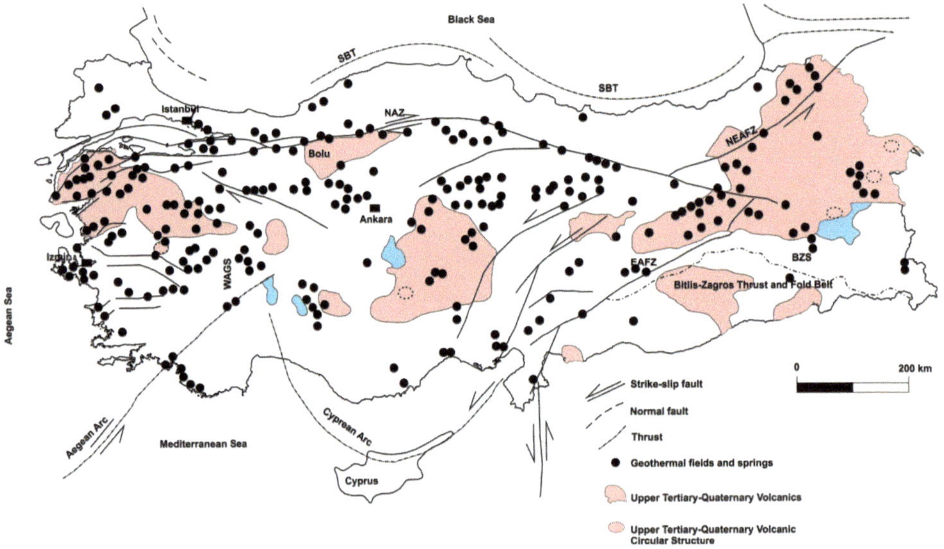

Figure 3.4 Geothermal sites in Turkey. The box represents the western Anatolian region (Chandrasekharam and Baba, 2022).

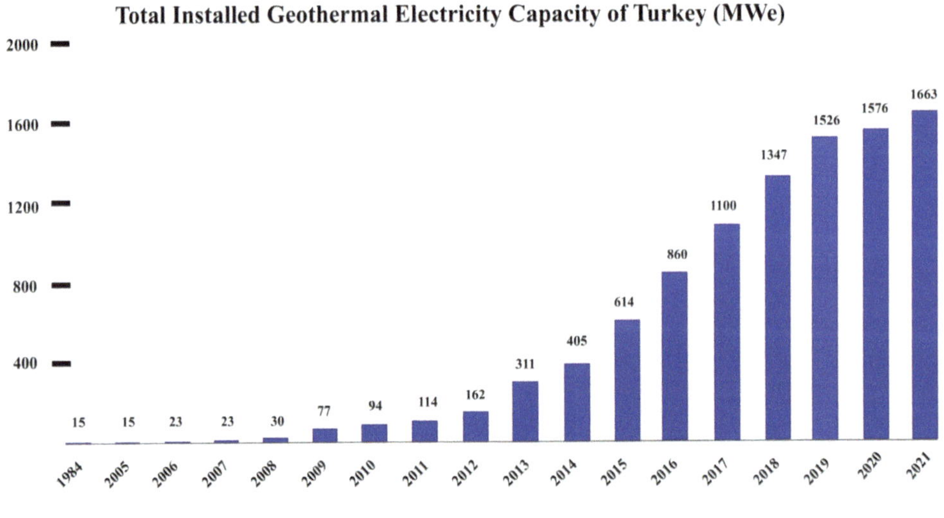

Figure 3.5 The installed capacity of geothermal power.

and Chandrasekharam, 2022), requiring 1 billion kWh of energy in a year. Thus, there is a huge amount of unutilized heat available to enhance direct applications. In addition to electricity supply and heat, the hydrothermal sources are contributing to reducing the carbon footprint of Turkey. Currently, CO_2 emissions from fossil fuels from electricity generation and heating homes are around 147 Mt (see Figure 3.2)

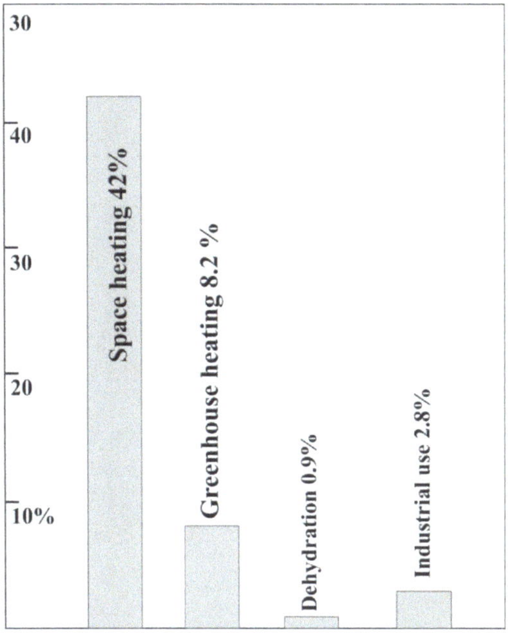

Figure 3.6 Direct application of geothermal energy status, Turkey.

(Chandrasekharam and Baba, 2022). These aspects are discussed in detail in the subsequent chapter.

While hydrothermal energy sources are site-specific, as seen from Figure 3.4, located along subduction zones, major rifts and rift basins and volcanic terrains, EGS has no such restrictions since granite intrusions are found at any place on the continents, and granites by nature are, in general, radiogenic due to the presence of radioactive elements like uranium, thorium and potassium, although their concentrations may vary. But certainly, granites do contribute to the heat flux on the surface of the earth. In certain countries, these granites occur in large quantities and with easy accessibility, while in other countries they are buried deep by a thick sedimentary cover. Turkey is one of the countries with huge radiogenic granite occurrences spread along western Anatolia (see Figure 3.3). The advantage is that they are easily accessible, have high radiogenic heat production and are located over a region of high heat flow and shallow Curie point depth (Aydin et al., 2019; Chandrasekharam and Baba, 2021, 2022). These characteristics have a great advantage for drilling operations and economics, which are discussed in the succeeding sections.

3.4 GEOLOGICAL AND TECTONIC EVOLUTION OF WESTERN ANATOLIA

While discussing the geological and tectonic evolution of western Anatolia, the focus of this chapter will be on the evolution of the high radiogenic granites of this region. The geological and tectonic fabric of western Anatolia (in fact the entire

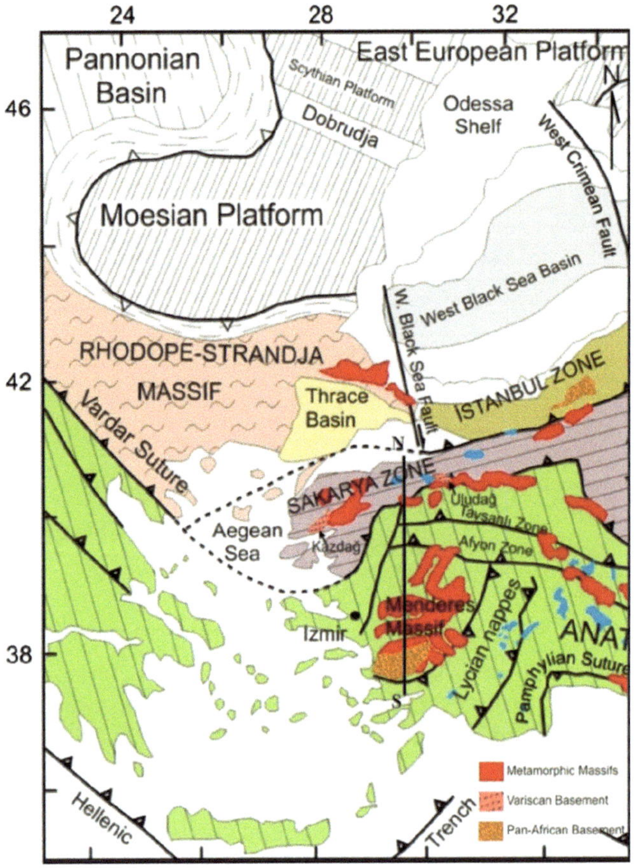

Figure 3.7 Tectonic fabric of western Anatolia (Gücer et al., 2016).

Turkey) shown in Figure 3.7, evolved due to the regional Alpine-Himalayan orogeny (Figure 3.8) and these tectonic elements are continuing to be active, like the Himalayas, due to northward compression of Turkey. Thus, the Sakarya zone (plate) is moving S-SW at the rate of 2 cm /year due to compression from the south and resistance from the north (Gurer et al., 2016). A regional synthesized general stratigraphic sequence of western Anatolia is shown in Figure 3.9. Some of the crustal rocks found in the higher Himalayas are represented in the SW and northern parts of the Western Anatolian region, like the Menderes Massif and the basement rocks of the Cretacous-Paleogene (Lower Tertiary) age in the northern part of the Sakarya zone. The sedimentary sequence representing the Paleogene and Neogene stratigraphy, that extensively outcrop in the western Anatolia, and those Mesozoic sediments, exposed toward the northern part of the western Anatolia, are intruded by granites of different ages that are shown in Figure 3.3. Some of them are 165 Ma (Nzegge, 2008) old, that are emplaced in the older sedimentary sequence, and others are of younger

Figure 3.8 The position of Turkey's structural fabric and the Himalayan-Alpine orogeny (https://web.itu.edu.tr/~okay/AralOkayMapsAndDiagrams.htm).

ages that are emplaced into the Paleogene and Neogene sedimentary sequence (Koksal et al., 2004; Gücer et al., 2016). The older rocks of granitic composition include monzonites, syenites and metagranites, that belong to the Alpine-Himalayanorogenic event, are dated Paleozoic age and occur north of the North Anatolian Fault (NAF) (Yilmaz, 1997; Bingöl et al., 1982). The older granitic intrusive, belonging to the post-collision tectonic event, are generated by partial melting of felsic granulites/tonalitic rocks that are abundant in the Precambrian crust of the Gondwanas, that was juxtaposed to the suture zone, an environment similar to the one that envisaged in the Himalayas (O'Brien, 2019). These Alpine orogenic granites are geochemically different from the Paleogene-Neogene granite intrusive that outcrop extensively along the western Anatolian region that are considered typically as arc-type plutonic members (Gorur et al., 1984; Tuysuz et al., 1994; Erdogan et al., 1996; Goncuoglu et al., 1997; Bingöl et al., 1982). A detailed description on the evolution of Cenozoic granites occurring in the western Anatolian region was given by Dilek and Altunkaynak (2007). The most characteristic feature of the younger and the older granite intrusive is their ability to generate high radiogenic heat due to the high concentration of uranium, thorium and potassium compared to other granites, detailed in the following section.

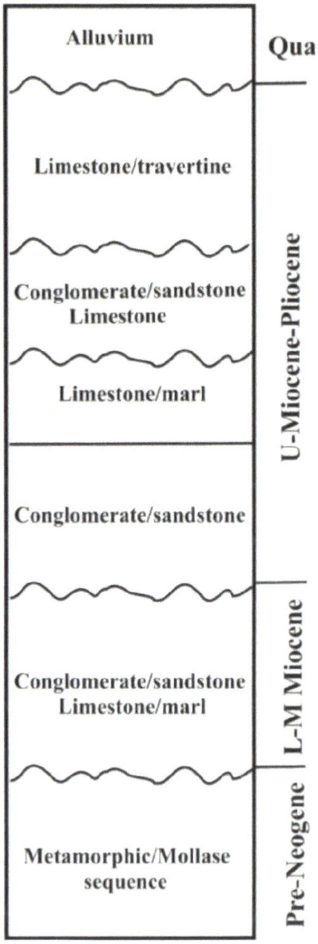

Figure 3.9 Regionally synthesized general stratigraphic sequence along western Anatolia. Adapted from Gurer et al., 2003.

3.5 RADIOGENIC GRANITES OF WESTERN ANATOLIA

The distribution of these granites in the western Anatolian region is shown in Figure 3.3. While the area occupied by these different age groups of granites is 6,910 km^2, the area occupied by the Cenozoic granites is much larger compared to other younger granites (Figure 3.3). The uranium, thorium and potassium content of the granites reported in the literature is compiled and shown in Table 3.2, along with the radioactive heat generation of the granites and the heat flow value resulting from these granites.

The radioactive heat generated by these granites (A) has been calculated following the equation of Rybach (1976)

$$A(\mu W/m^3) = 10^{-2} * \rho * (9.52_C U + 2.56_C Th + 3.48_C K),$$

where ρ (g/cm^3) is the specific gravity of the granite (2.7 g/cm^3), and $_C U$, $_C Th$ and $_C K$ are the concentrations of uranium (ppm), thorium (ppm) and potassium (%), respectively.

Table 3.2 Concentration of U, Th and K, radioactive heat generation (RHP) and heat flow values of granites of western Anatolia

Pluton name	*Area* (km^2)	*Th (ppm)*	*U (ppm)*	*K (%)*	*Radioactive heat production* ($\mu W/m^3$)	*Heat flow* (mW/m^2)	*References*
Gördes	19.56	29.6	45	1.43	13.75	177.47	Erkül and Erkül (2012), Esenli et al. (2019)
Alaçamdağ	165.59	32	10	3.4	5.10	91.02	Akay (2008), Erkül and Erkül (2012)
Koyunoba	128.28	30.4	4.8	4	3.71	77.11	Akay (2008), Erkül and Erkül (2012)
Eğrigöz	565.82	51	9.1	3.3	6.17	101.74	Ozgenc and Ilbeyli (2008), Akay (2008), Erkül and Erkül (2012)
Salihli	31.11	14.9	20.6	3.5	6.65	106.54	Catlos et al. (2011), Dilek (2009), Erkül and Erkül (2012)
Turgutlu	2.41	15.3	4.5	4.6	2.65	66.46	Catlos et al. (2011), Dilek (2009), Erkül and Erkül (2012)
Yatağan	100.00	10.8	22.82	4.53	7.04	110.38	Bozkurt et al. (2006)
Bafa	100.00	4.1	40.1	4.3	10.99	149.95	Erdoğan and Güngör (2004)
Orhaneli	447.69	46.1	30.4	4.64	11.44	154.36	Delaloye and Bingöl (2000), Altunkaynak et al. (2012)
Kaymaz	7.81	59.4	26.8	4.61	11.43	154.28	Shin et al. (2013), Örgun et al. (2005)
Topkaya	50.52	20.4	5.7	2.2	3.08	70.82	Demirbilek (2012)
Günyüzü	99.11	14.2	4.5	4.24	2.54	65.37	Demirbilek et al. (2018), Bağcı et al. (2019)
Kızıldam	23.05	40	13.5	2.87	6.50	105.05	Delaloye and Bingöl (2000), Karacık et al. (2008)
Çamyayla	36.84	79	21	3.49	11.19	151.86	Aydin et al. (2019)
Kuşçayır	26.40	107	22	2.24	13.26	172.61	Aydin et al. (2019)
Evciler	272.33	47.4	8.4	3.56	5.77	97.70	Erdoğan et al. (2013), Birkle (1992), Genc (1998), Okay and Satır (2000), Özdamar (2018)
Yolindi	102.47	14.4	3.4	2.51	2.11	61.05	Aysal et al. (2012)
Sarıoluk	30.00	33.4	11	2.9	5.41	94.09	Karacık et al. (2008)
Çataldağ	673.67	10.3	16.3	3.22	5.20	92.04	Boztuğ et al. (2009), Delaloye and Bingöl (2000)
Eybek	212.78	31	9.7	2.8	4.90	88.99	Black (2012), Delaloye and Bingöl (2000)

(Continued)

Table 3.2 (Continued) Concentration of U, Th and K, radioactive heat generation (RHP) and heat flow values of granites of western Anatolia

Pluton name	*Area (km^2)*	*Th (ppm)*	*U (ppm)*	*K (%)*	*Radioactive heat production ($\mu W/m^3$)*	*Heat flow (mW/ m^2)*	*References*
Uludağ	770.02	15.6	10.4	2.7	4.01	80.05	Topuz and Okay (2017), Delaloye and Bingöl (2000)
Ilıca	247.57	25.1	7.4	2.3	3.85	78.53	Boztuğ et al. (2009), Karacık et al. (2008)
Kozak	825.10	22.5	5.4	2.93	3.22	72.19	Delaloye and Bingöl (2000), Black et al. (2012), Boztuğ et al. (2009)
Kapıdağ	217.47	30.85	5.19	4.44	3.88	78.84	Karacık et al. (2008)
Avşa	37.74	17.17	5.78	2.93	2.95	69.48	Karacık et al. (2008)
Karabiga	126.55	29.06	2.68	3.94	3.07	70.68	Karacık et al. (2008)
Şevketiye	44.80	34.9	11.9	2.96	5.75	97.49	Karacık et al. (2008)
Çamlık	77.93	12.5	4.4	2.22	2.20	62.04	Aysal et al. (2012)
Yenice	97.51	35.7	9.6	3	5.22	92.17	Karacık et al. (2008)
Karacabey	22.55	9.7	2.9	0.87	1.50	54.98	Sunal (2013)
Baklan	16.30	42.9	14.1	2.6	6.83	108.34	Aydoğan et al. (2008), Akay (2008), Erkül and Erkül (2012)
Kestanbol	219.13	80	17.4	3.98	10.38	143.76	Black et al. (2012), Akal (2013), Delaloye and Bingöl (2000), Şahin et al. (2010), Chandrasekharam and Baba (2021)
Pamukova	46.06	21.3	3.85	0.61	2.52	65.19	Çelik et al. (2009)
Sarıcakaya	582.08	22.3	7.3	4.37	3.83	78.28	Othman (2016)
Fıstıklı	102.97	17.6	6.9	1.24	3.11	71.07	Altunkaynak et al. (2012)
Esenköy	58.85	20.07	2.81	3.85	2.47	64.71	Yılmaz et al. (2010)
Beypazarı	321.97	42.8	5.3	3.39	4.64	86.39	Speciale et al. (2012)

The location of the granites listed here are shown in Figure 3.3. The total area of the granite outcrop is 6,910 km^2. The concentration of radioactive elements is taken from the published work listed under references.

$_C$K is expressed as a percentage of elemental potassium (or K_2O multiplied by 0.83). The heat flow values were calculated using the equation (Lachenbruch, 1968).

$$Q = Q_0 + D * A$$

where Q is the heat flow at the surface, Q_0 is an initial value of heat flow unrelated to the specific decay of radioactive element at a certain time, D is the thickness of rock over which the distribution of radioactive element is just about homogeneous and A is the radioactive heat production. The crustal thickness along western Anatolia being ~25 km (Tezel et al., 2013), the natural heat flow value (emanating from the mantle) of 40 mW/m^2 (Tezel et al., 2013) is considered in the above equation. Except for a few, the RHP of the granites is above the average heat generation of the continental crust (5 μW/m^3, Rybach, 1976). Further, the heat flow value calculated for these granites is 68–107 mW/m^2 (Table 3.2), and this value concurs with the heat flow value reported for this region based on Curie point depth deduced from the aeromagnetic investigation (110 mW/m^2, Aydin et al., 2019) and concurs with heat flow values deduced from the helium isotopic systematics in geothermal fluids in western Anatolia (Gulec and Hilton, 2006).

The Curie point depth temperature recorded from the aeromagnetic survey is shown in Figure 3.10, and the heat flow values deduced from this observation are shown in Figure 3.11.

The heat flow values observed over western Anatolia are 50% higher than the world average (Ilkisi and Yenigun, 1997). In fact, the highest heat flow values have been recorded from Gediz and Buyuk Menderes grabens (near Menderes Massif, see Figure 3.12) with an average value of 120 mW/m^2 and a recorded temperature of 450°C at about 10 km depth, which is close to the Curie point depth reported

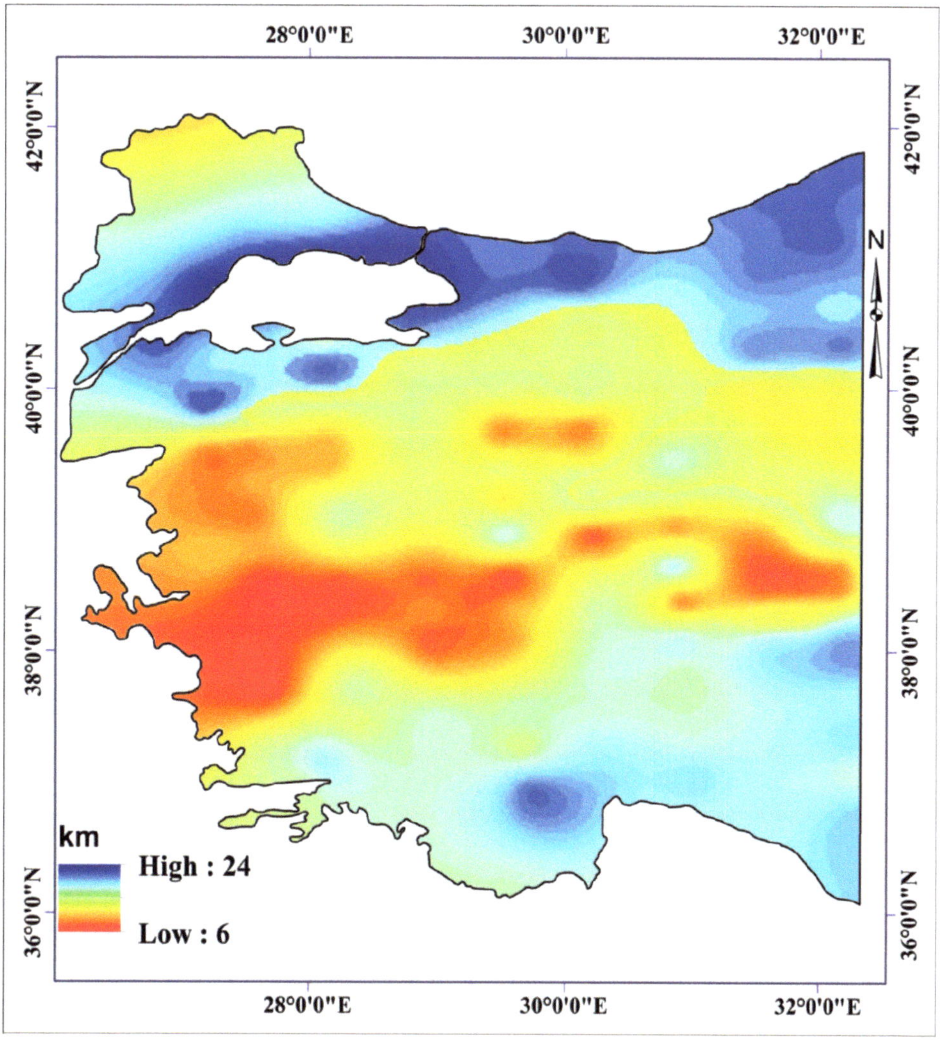

Figure 3.10 Curie point of depth deduced from an aeromagnetic survey. Modified after Aydin et al., 2019.

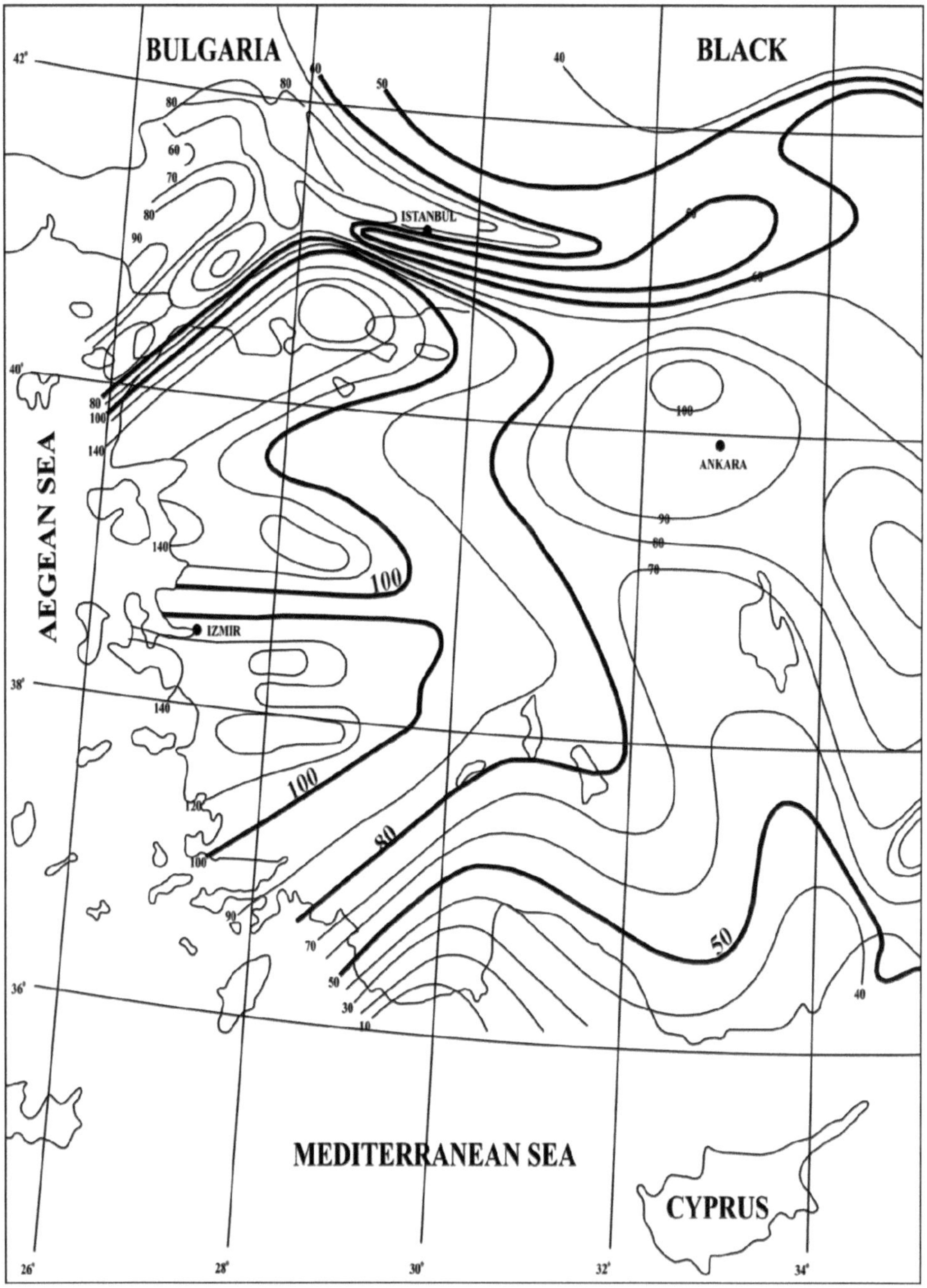

Figure 3.11 Heat flow values over the western Anatolian region were deduced from the Curie point of depth. Modified after Aydin et al., 2019.

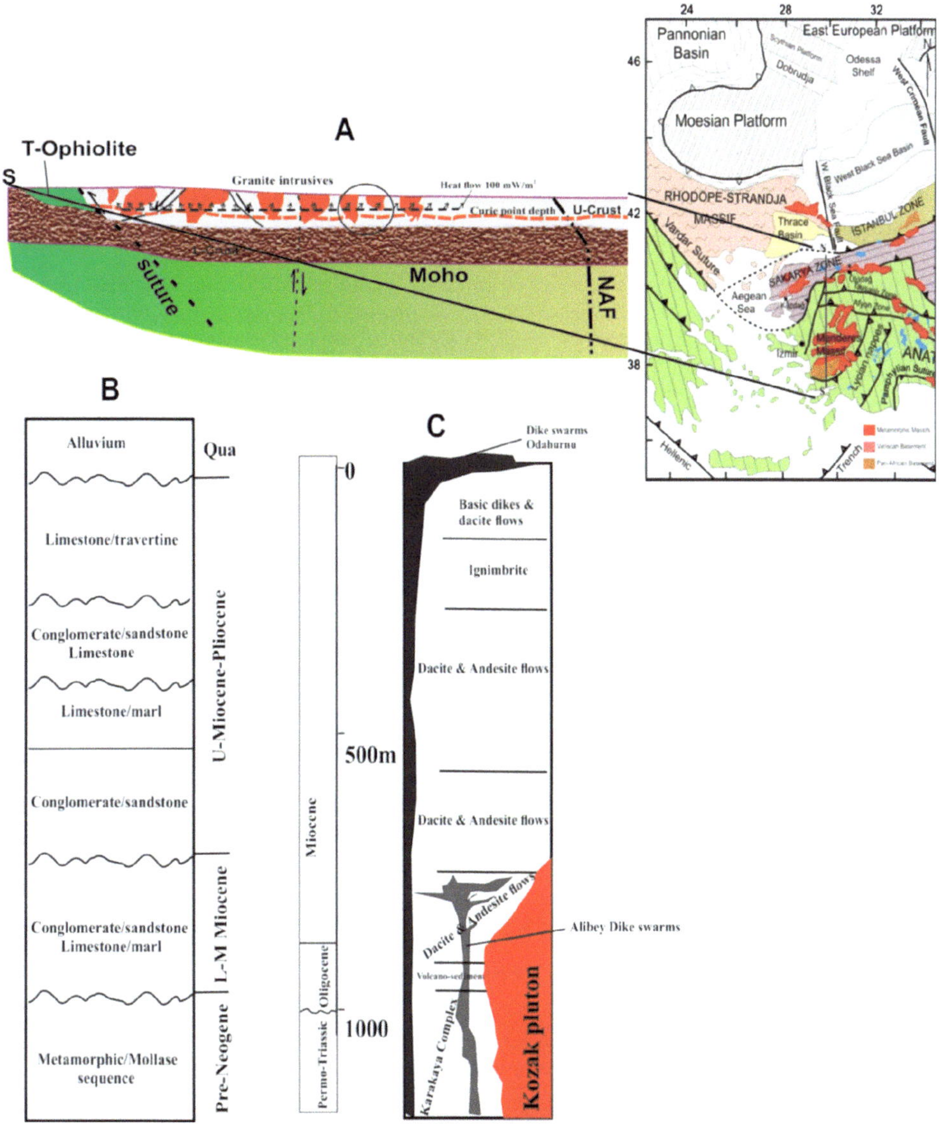

Figure 3.12 Schematic cross-section along line N-S shown in Figure 3.7 and the sedimentary cover over the granites in Kozak (B) and other intrusive granites (C). The Late Oligocene–Early Miocene granites have intruded into the older and younger volcanic flows and are covered by thick post-Oligocene sediments at several places in western Anatolia. The circled area in this figure is enlarged and shown in Figure 3.13, which represents a schematic cross-section across the line C-D in the insert map. Adapted from Aldanmaz et al., 2000; Tezel et al., 2013; Gurer et al., 2001, 2003; Aydin et al., 2019; Altunkaynak et al., 2012; Dilek et al., 2007; Bingöl et al., 1982; Ilkisik et al., 1997; Bozkurt and Oberhansli, 2001.

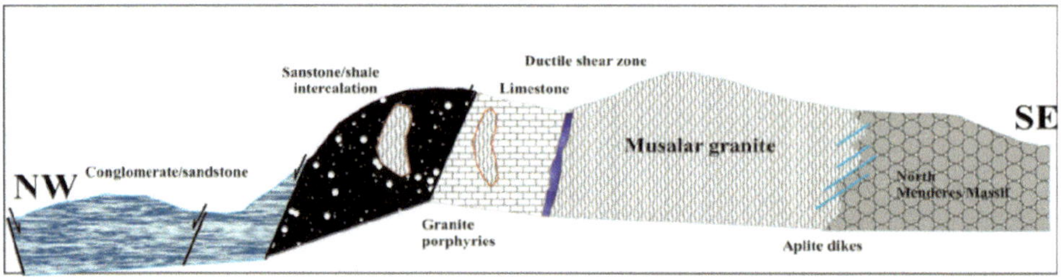

Figure 3.13 Schematic cross-section along the C-D line (Figure 3.12) showing the lithological and tectonic features relevant to EGS projects. Adapted from Chandrasekharam et al., 2022; Erkul and Erkul, 2010. Different available options to extract the heat from the granite are explained in Figures 3.17a–d (in Section 3.9), taking the area circled as a reference.

based on the aeromagnetic data. A schematic section showing the tectonic regime, magmatic activities (especially granitic intrusive) and thermal regime along the N-S direction of western Anatolia is shown in Figure 3.12. This section is developed based on the data from Dilek and Altunkaynak (2007), Aydin et al. (2019) and Chandrasekharam and Baba (2021). In fact, a conductive layer below the Menderes Massif, deduced from the MT (Magneto-Telluric) survey, is identified at 10 km, which concurs with the Curie point depth and indicates the zone of crustal melting in this region (conductive layer curve at this depth fits well with the curve of wet granite melting at heat flow >85 mW/m^2, Adam, 1987). Considering the coincidence of a conductive layer, high heat flow, shallow earthquake focal depths and shallow Curie point depth in this region, the presence of a ductile and partially melted (visco-elastic) lower crust cannot be ruled out in this region (Gurer et al., 2001). A cross-section across line C-D in the insert map in Figure 3.12 is shown in Figure 3.13, which gives an idea about the subsurface lithological configurations in the circled area in Figure 3.12.

3.6 STRESS REGIME OF WESTERN ANATOLIA

The Mediterranean region, including the Aegean Sea, exhibits complex regional dynamic stress field patterns due to the convergence of African and Eurasian plates and the involvement of microblocks of different sizes, strengths and geometry (Rebai et al., 1992). Therefore, the state of stress depends on these geodynamic states. Thinned continental crustal zones are, like the Aegean back-arc basin, under an extensional tectonic regime. The internal deformation of the microblocks results in stress deviations at different geological time scales, as observed within the Turkey microplate. The Aegean region, in general, is dominated by strike-slip and extensional tectonics, as stated above, due to the northward movement of Africa (1 cm/y) and Arabia (3 cm/y) relative to Eurasia (Rebai et al., 1992; Jackson, 1994). The net dynamic stress due to the megaplate convergence resulted in a westward slip of the Turkey microplate along the North and East Anatolian Faults (Figure 3.14). The crustal extension is accommodated by the Hellenic trench. Fundamental contributions to understanding the Aegean tectonic regime have been made by several authors. According to McKenzie (1972), the

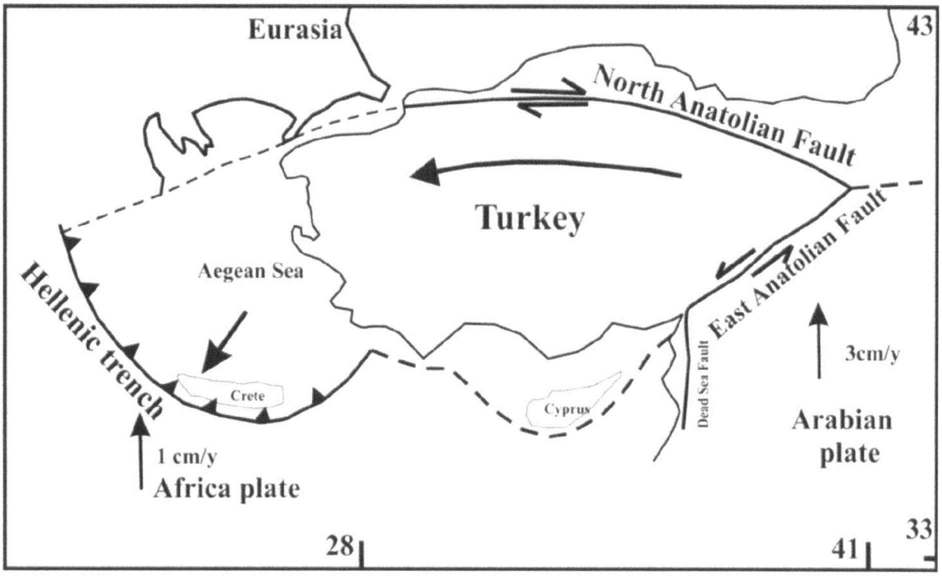

Figure 3.14 Map showing the westward movement of Turkey relative to Eurasia, accommodated by the strike-slip faulting of the North (NAF) and East Anatolian faults (EAF) and crustal shortening in the Hellenic trench, a junction between the northward moving African plate and the SW movement of Turkey. The slip rate of the NAF is 36 mm/y, while the slip rate of the EAF is 33 mm/y. Adapted from Jackson, 1994.

present-day motions in the Aegean regions (the western Turkey and eastern Aegean region) should be viewed in conjunction with the tectonic movements that are taking place in eastern Turkey. The crystal dynamics over Turkey, resulting from the above continental collision caused crustal thinning, normal faulting (horst and graben structures, pull-apart basins) and the formation of sedimentary basins/graben structures such as the north and south Menderes graben, Denizli basin, Gediz graben, Simav graben and Bakırçay graben (Bozkurt and Mittwede, 2005; Uzel et al., 2012). The high thermal regime recorded over western Anatolia, in the form of high heat flow, a high geothermal gradient and shallow Curie point depth, as described above, is the result of crustal thinning and intrusive magmatic activity. Magmatic activity is represented by several granitic intrusions and volcanic eruptions. The occurrence of granitic intrusion over a wide geological time scale (Eocene-Miocene) and the occurrence of earthquakes indicate an active tectono-magmatic environment over the entire eastern Aegean region (Jackson, 1994; Uzel et al., 2012; Ozden et al., 2018; Sangu et al., 2020; Gurer et al., 2013, 2021).

The major regional tectonic factors (Figure 3.14) have given rise to a large number of ancillary tectonic fabrics in the form of active faults, suture zones and satellite faults around the major thrusts and slips that are observed in the formation outcrops (like those seen in Figures 3.4 and 3.7) (Dilek et al., 2009; Gurer et al., 2021; Gorur et al., 1984; Ocakoglu et al., 2004; Okay and Tuysuz, 1999), like the Vardar-Izmir-Ankara suture (Figure 3.7) that represents the hanging wall of the Cretaceous to Paleogene subduction accretion complexes (Okay and Tuysuz, 1999; Okay, 2011; Gessner et al., 2016). These active tectonic movements have induced an extension and compression regime over the entire western Anatolia since the beginning of the collision between the Arabian, African and Eurasian plates within the framework of the Himalayan-Alpine orogeny.

Thus, Turkey is a very complex tectonic zone and is the most seismically active zone due to the geodynamic activity controlled by the collision between the Eurasia-Africa-Arabia plate. The compressional deformation due to the northward movement of the Arabian plate is causing westward extrusion of the Anatolian block along the NAF and EAF, resulting in the thickening of the crust on the eastern side of Turkey. The northward subduction of the African plate beneath the Eurasian plate along the Hellenic arc is responsible for the observed tectonic mosaic over the entire western Anatolian region. This process is responsible for the extensional tectonics over the western Anatolian region (McKenzie, 1972; Bozkurt, 2001; Le Pichon and Angelier, 1979; Jackson and McKenzie, 1988; Kissel and Laj, 1988; Meulenkamp et al., 1988; Avigrad et al., 1997). As an example, it is focused on the development of the Biga peninsula, a part of the western Anatolian tectonic regime that encloses several highly radiogenic granitic intrusives (Figure 3.7), especially the Kestanbol plutons (Figure 3.7, Table 3.1). The Biga peninsula developed a complex pattern of deformation due to the intersection of the NAF system and the NW-SE extension fabric, represented by shear zones, pull-apart or transitional pull-apart basins, and regional fault, graben and horst systems (Figure 3.8) during the Middle Pleistocene (Şengör et al., 2005; Gürbüz and Gürer 2009, Yılmaz, 2017; Sangu et al., 2019, Taymaz et al., 1991, 2017; Yılmaz et al., 2000; Bozkurt, 2001, Gürer et al., 2006, 2016).

In order to assess the geodynamic status of the western Anatolian region, extensive structural fieldwork has been conducted by Gurer et al (2021) in and around the Bega peninsular region. Nearly 240 fault slip data points (dip amount, dip direction, fault plane, azimuth and plunge striation and slickenside direction) were analysed from fresh rock exposures. The paleostress analysis was carried out using the Direct Inversion Method (INVID) of Angelier (1990) in conjunction with SG2PS software developed by Sasvari and Baharev (2014). In geodynamics (the notations for stress axes are: σ_1 (maximum stress), σ_2 medium stress and σ_3 minimum stress directions), the stress ratio R is represented by ($\sigma_2-\sigma_3/\sigma_1-\sigma_3$) (Angelier, 1994). Two compressional regimes were reported in this region: an initial NE-SW compression followed by NW-SE compression. The NE-SW compression (Sh_{max}) was the dominant force during the Miocenes, and the NW-SE (Sh_{max}) was the dominant force during the Late Pleistocene-Holocene period. This indicates an anti-clockwise rotation of the stress maximum. The NW-SE Sh_{max} appears to be the dominant stress in the present day, as indicated by the 2017 earthquake in this region (Özden et al., 2018).

In order to investigate the imprint of the major regional tectonic fabric on the rock microfabric, we collected field-oriented Oligocene-Miocene granite samples from Kozak (Figure 3.8) within the Biga peninsula to study the microfabric structure in the granite. Polysections from the rock sample were prepared to examine the surface phenomena on the grains of the minerals (especially quartz, since it lacks any cleavage) under the scanning electron. Quartz grains were examined, and the fractures in the quartz grains were mapped. The SEM photo is shown in Figure 3.15.

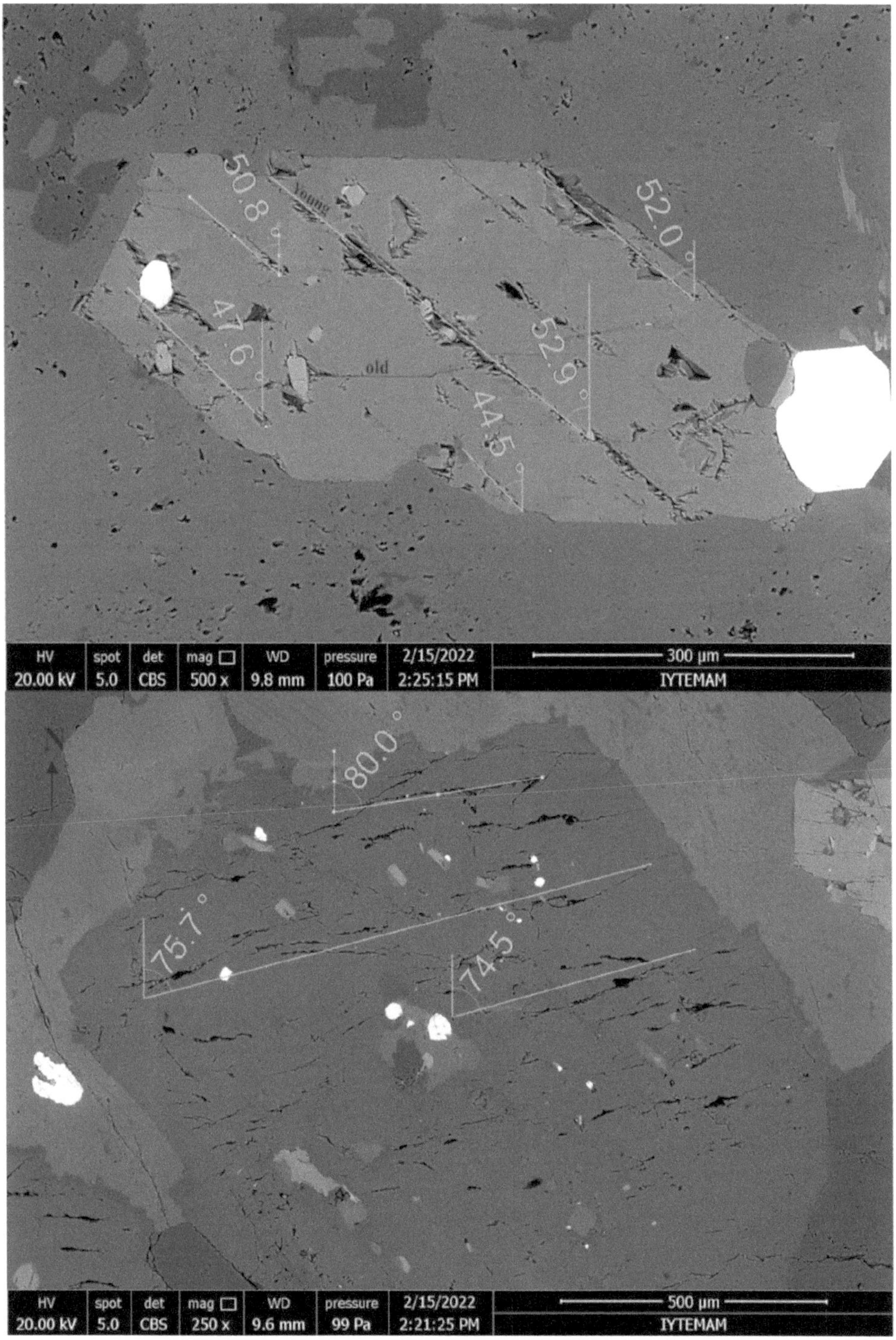

Figure 3.15 SEM photographs of Kozak granite showing stress patterns in quartz.

Two sets of fractures are observed in the quartz grains under the SEM. The NE-SW cracks appear to be older (Figure 3.15a). From Figure 3.15a, it appears the granite was under a NE-SW compressional regime during the Oligocene period, when it was emplaced into the upper crustal layers (Figure 3.12). Whether these fractures are related to the regional tectonic fabric or not, what is clear is that they provide information on the existence of fractures in the granites and their orientation. This will help to a great extent during the planning stages of hydrofracturing to create a fracture network in Kozak granite.

3.7 POWER GENERATION CAPACITY OF THE GRANITES

Granites with RHP capacity are being exploited for electricity generation. Some of the projects that are live are listed in Table 3.3.

For EGS to be successful, as described above, a hot granite with sufficient heat generation, a thermal insulation cover located over an area of high hat flow, and a convenient stress regime are essential (Chandrasekaram et al., 2016b; Chandrasekharam and Baba, 2021; Chandrasekharam et al., 2015). The Kozak and Kestanbol granites meet these criteria and are hence favourable sites in the western Anatolian region for initiating EGS projects. As indicated above, the geothermal gradient in this region is above 60°C/km, and hence these two granites, with their inherent high heat capacity, should be able to attain 180°C at 3 km depth. Based on previous EGS projects, it is estimated that 1 km^3 of such hot granites can generate 79×10^6 kWh of electricity for 30 years (Somerville et al., 1994; Cooper et al., 2010; Chandrasekharam et al., 2015). Based on the previous EGS projects, especially the hot dry rock project of Cooper Basin in Australia, a relationship between hat flow values and levelized cost of power was established (Somerville et al. 1994) as shown in Figure 3.16.

Table 3.3 EGS projects in operation

EGS project	*Power plant type*	*Fow rate (L/S)*	*Installed electrical*	*Thermal*
			Capacity: MWe	Capacity: MWt
Landau, Germany[a]	Organic Rankine Cycle	80	3.6	5
Insheim, Germny[a]	Organic Rankine Cycle	85	4.8	10
Unterhaching, Germany[a]	Kalina Cycle	150	3.3	38
Soultz, France[a]	Organic Rankine Cycle	30	1.5	N
Altheim, Germany	Organic Rankine Cycle	81	1	12
Hijori	Organic Rankine Cycle	17	0.13	8
UDDGP, UK[b]	Organic Rankine Cycle	60	3	30

Source: Adapted from Baba and Chandrasekharam, 2022.
UDDGP, United Down Deep Geothermal Project.
[a]Power plants in operation.
[b]Power plants under construction.

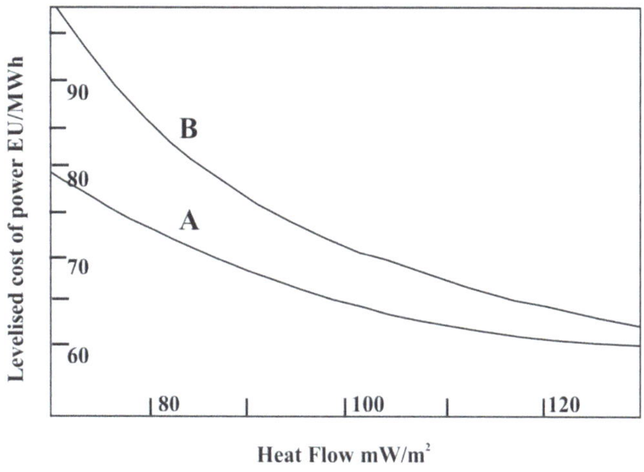

Figure 3.16 Levelized cost of power from EGS sources (a) 180 mW/m²; (b) 160 mW/m². Adopted from Cooper et al., 2010.

3.8 ADVANTAGES OF EGS

3.8.1 The cost factor

Unlike the hydrothermal system, the EGS is in its development stage; hence, the electricity tariff and cost of heat are estimated based on the heat in place in the granite at a certain depth, the heat flow values (Figure 3.11) and the geothermal gradient (Cooper et al., 2010). In the case of the granites of western Anatolia, these highly radiogenic granites (Figure 3.3, Table 3.1) are placed over a region of high heat flow, and in addition, the Curie point of depth is very shallow (Figure 3.10) over the entire western Anatolia. In addition, the depth at which high temperatures can be located in the granites is shallow, minimizing the cost of drilling to harness the heat (Baumgartner et al., 1995). These factors further reduce the levelized cost of electricity below 6 euro cents (Table 3.4, Figure 3.16). Although the energy cost of fossil fuels is much lower compared to geothermal (Table 3.4), the cost of mitigating environmental factors caused by the fossil fuel-based energy sources is much higher (i.e., cleaning the environment due to high emissions, disposing of fly ash from the thermal power plants, mitigating issues related to groundwater pollution in and around thermal power plants and health hazards, etc.) than the energy cost itself. There are no such hidden costs in geothermal energy sources. The factors related to CO_2 emissions are discussed separately in the following section.

Turkey's electricity demand is projected to grow beyond 93% by the year 2040 (Diddglio et al., 2020). To meet this demand, Turkey needs nearly 9 billion euros per year. This is greater than what is expected if the prevailing energy supply policy is continued, i.e., using fossil fuels as the primary energy source without inducting another energy source like geothermal into the primary energy mix. This is known as

Table 3.4 Levelized cost of electricity from fossil fuels and geothermal energy sources

Source	*Capacity factor*	*Fixed O & M Euro/MWh*	*Variable O & M Euro/MWh*	*Levelized cost/euro cents/kWh*
Coal	85	3.5	24.7	7
Gas	30	2.2	58	2.6
Hydrothermal	92	10.24	0	8.97
EGS	92	9.39	0	6–9

Source: Modified after Baba and Chandrasekharam, 2022.

business as usual attitude (BAU). This BAU will increase the emissions of CO_2 by over 30% (Diddglio et al., 2020). Thus, the country will incur additional expenditures to clean the environment or implement policies to mitigate CO_2 emissions. These policies could be implemented by buying carbon credits from countries emitting low carbon or by imposing carbon taxes (like European countries, Diddglio et al., 2020). All this will add up to escalating the consumer price of electricity. To avoid such situations, the country should take advantage of its huge EGS source available at an affordable cost along the western Anatolian region (Baba and Chandrasekharam, 2022; Chandrasekharam and Baba, 2022).

3.8.2 CO_2 emissions status

Turkey's annual electricity consumption is about 384 billion kWh, and fossil fuels, as described above, are the primary energy source (Table 3.1). The per-capita electricity consumption is about 4,660 kWh per year (Diddglio et al, 2020; IEA, 2021b; Chandrasekharam and Baba, 2022). According to the projections, if the policy is to follow the BAU attitude, i.e., not make any changes in the energy sector to reduce carbon emissions, then fossil fuels will continue to be the major contributor to the primary energy source (Keles and Bilgen, 2012; Diddglio et al., 2020; IEA, 2021a; Chandrasekharam and Baba, 2022). With energy demand expected to grow beyond 70 Mtoe in the coming decade from the current 58 Mtoe (Table 3.1), without any major low-carbon energy source contributing to the primary energy mix (IEA, 2021c), Turkey's carbon emissions will surpass 500 Mt in the next decade from the current 400 Mt (Yelden and Voyvoda, 2015; Ritchie and Roser, 2020) due to growth in population (Figure 3.1). Next to the power sector, the space heating and cement industries contribute 147 and 135 Mt CO_2 to the atmosphere, respectively (Figure 3.2) (Chandrasekharam and Baba, 2022). Turkey has agreed to implement the Paris 2016 agreement, reduce CO_2 emissions and meet the 2D scenario recommended by the IPCC (IPCC, 2018) by reducing emissions from the current 500 to 390 Mt (Figure 3.1) (Chandrasekharam and Baba, 2022). In order to achieve this status, the country should invest in the technological development of innovative renewable energy sources. The hydrothermal energy sources are already generating a substantial amount of electricity (Figure 3.3, 1,663 MWe) (Baba et al., 2021; Baba and Chandrasekharam, 2022; Chandrasekharam and Baba, 2022), thus contributing to the emissions reduction strategy. However, this source has not been exploited to its fullest capacity (Baba and Chandrasekharam, 2022). The best option

Turkey has for implementing sustainable economic growth and contributing to the CO_2 reduction programme is to enhance the utilization of its hydrothermal energy resources and unlock the energy potential of the highly radiogenic granites spread over an area of 6,910 km^2 (Figure 3.3, Table 3.1) along western Anatolia. The energy that can be harnessed from these granites is quite substantial. For example, a 10-km-thick granite body from Kestanbol, and its minimum area of 220 km^2 (Figure 3.3, Table 3.1), with its RHP of 7 $\mu W/m^3$ (Table 3.1), can generate 176 billion kWh of electricity. Similarly, a 10 km thick Kozak pluton with its RHP of 7 $\mu W/m^3$, can generate 658 billion kWh of electricity (Chandrasekharam and Baba, 2021, 2022; Baba and Chandrasekharam, 2022). Thus, a minimum amount of 546×10^9 kWh can be generated from 6,910 km^3 of granite from western Anatolia. Although it may not be possible to mine the entire heat available within the granite body, at least 50% of the heat can be mined. This amounts to nearly half the current power demand in Turkey. Thus, EGS sources will help the country to comply with the carbon reduction policy and bend the carbon index curve, if not to the 2D scenario of the IPCC (IPCC, 2018), at least towards it (Chandrasekharam and Baba, 2022).

3.9 DISCUSSIONS

The western part of Turkey has an unusual tectonic fabric and a complex magmatic evolutionary history that cannot be found anywhere else in the world. The region has undergone multiple stages of deformation due to dilation and contraction of the crust, resulting in the formation of pull-apart basins, normal thrust faulting, subduction processes and concurrent magmatic activity represented by volcanic eruptions and granitic intrusions (see Section 3.6). Thinning of the crust and foundering of the crust have enabled high-temperature isotherms to rise to shallower levels, thereby creating a high-temperature regime at shallower crustal levels (Figure 3.12) and enabling the formation of huge geothermal reservoirs along the western margin of the country. The involvement of the mantle in the evolution of the hydrothermal system is evident from the high $^3He/^4He$ ratio and high R values (Gulec and Hilton, 2006). Nearly 10%–30% of the mantle He signature is found in all the geothermal waters occurring in the western Anatolian region. The R (R/R_A: ratio of $^3He/^4He$ in the sample (R) to the value of in the air R_A) values in the thermal waters vary from 0.1 to 3.68. The R-value of the crust is 0.02 and that of the mantle is 8 (O'Nions and Oxburgh, 1988; Giggenbach et al., 1993; Gulec and Hilton, 2006). Thus, the plutons of granitic composition (Figure 3.3), spread over an area of 6,910 km^2, are emplaced into a dynamic tectonic and high thermal regime environment (Figures 3.10 and 3.11). The area north of the Izmir-Anatolia fault zone is the locus of the Cenozoic granitic intrusive, while the area south of this fault zone is occupied by the Precambrian granitic intrusive (Figure 3.3). The stratigraphic section of Kozak pluton (Figure 3.12) shows that the rock is thermally insulated by thick volcanic and sedimentary layers and is partly exposed to the surface, unlike the high radiogenic granites of Slutz, France (Chapter 2) this volume) and Cooper Basin, Australia (Somerville et al., 1994) where hot granite is located 5 km deep. For any simulation or modelling, this large depth has to be considered. A schematic cross-section drawn across the Biga peninsula (Figure 3.13) gives a prospective idea about the lithology and fracture network that will help to establish an EGS project in western Anatolia in terms of numerical modelling and investment opportunity. The most striking feature is the presence of a shear zone at the contact between the granite and the limestone and fractures extending between the sedimentary sequence and the granite. In fact, the United Down Deep Geothermal Project (UDDGP) project (Table 3.2) targeted the shear zone in the granite to establish a non-conventional method of extracting heat from the granite (Cotton et al., 2018). Further, the surface tectonic framework and the stress pattern discussed in Section 3.6 (Figures 3.14 and 3.15) clearly demonstrate the pattern of heat exchanger that can be induced in the granite at 3 km depth. Now that directional drilling technology is maturing (Cotton et al., 2018), the western Anatolian granite terrain provides excellent sites for implementing EGS projects with lower capital expenditure and lower power tariffs. The amount of power and heat that can be generated (Section 3.7) will, to a large extent, offset the use of fossil fuels as a primary source, and Turkey can plan its strategy for sustainable development by reducing fossil fuel imports and mitigating CO_2 emissions in the coming decade. The technology options that can be adopted to mine the heat are shown in Figures 3.17–3.20. It has been used in the subsurface section shown in Figure 3.13 to discuss the possible options available to harness the heat from the granite:

Option 1: Conventional hydrofracturing of the granite (Figure 3.17):

This technology is traditional in that it induces a fracture network in the granites at a certain depth, and the fracture propagation is controlled by the stress regime acting on the rock. This option was implemented during the initial EGS development in Slutz, France (see Chapter 2, Cooper Basin, Australia (Somerville et al., 1994; Cooper et al., 2010)).

Option 2: (Figure 3.18)

This option is cost-effective since drilling into the sedimentary rocks is much easier compared to drilling into the hard rock. The heat exchanger has no direction here, and a large network of fractures connecting the two wells should be established.

Option 3: (Figure 3.19)

This option is similar to the one being implemented in the UDDGP (United Downs Deep Geothermal Power, United Downs Industrial Estate site in Carnwall, UK) EGS

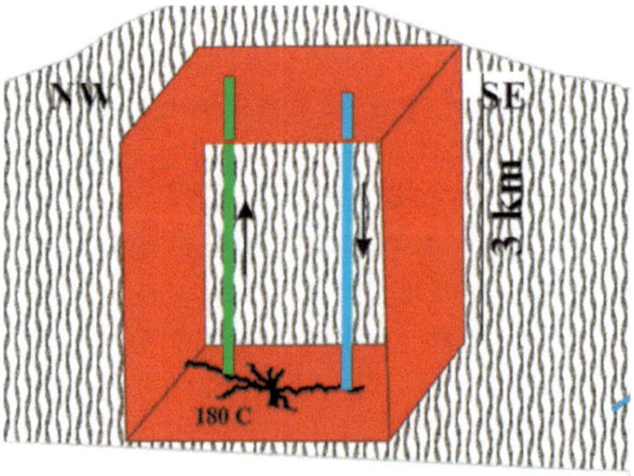

Figure 3.17 Schematic diagram showing the conventional EGS technology where the injection and production wells are connected through a network of fractures in the granite body (Musalar granite: see the circled area in Figure 3.13. Here the SH_{max} is NW-SE) (see Section 3.6 and Figure 3.15).

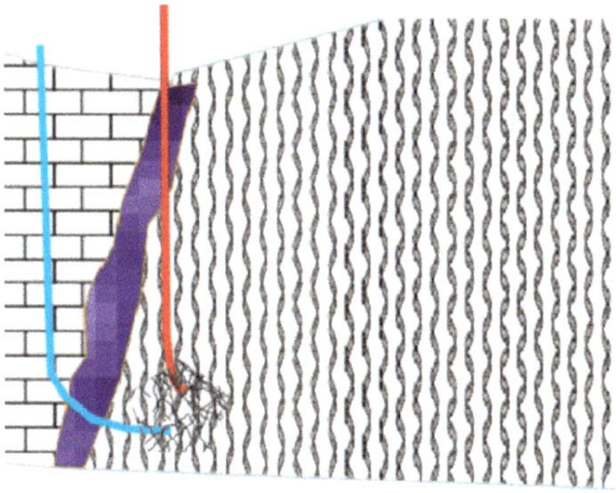

Figure 3.18 The cost of drilling is reduced due to the directional drilling (injection well) through the limestone formation into the hot granite at 3 km depth. A fracture "mesh" can be induced connecting the injection and production wells. The production well also penetrates only partially into the hard rock, thus reducing the cost of drilling. For details of the rock types, see Figure 3.13 (Musalar granite: see circled area in Figure 3.13).

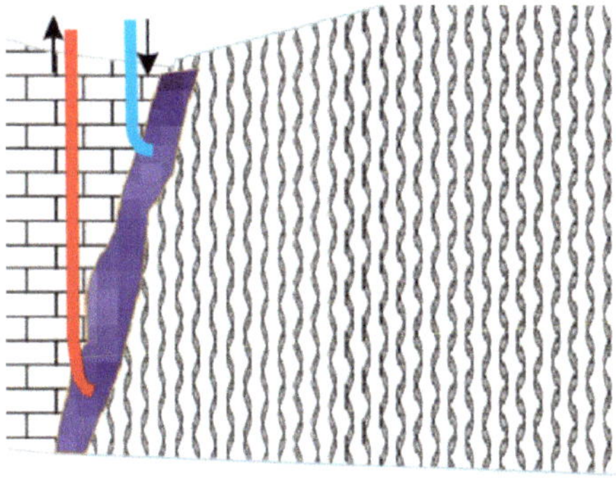

Figure 3.19 Schematic diagram showing the advantage of shear in harnessing heat using EGS technology. For rock types, see Figure 3.13 (Musalar granite: see circled area in Figure 3.13).

project in Cornwall, UK (Cotton et al., 2018). In the UDDGP project, which targeted a natural shear zone, the shallow injection wells and the deep production well are separated by about 2 km, and the production well drilled to 4.5 km registered a bottom hole temperature of 190°C. The hot water at the surface registered 175°C. This project is expected to generate 1–3 MWe (Cotton et al., 2018). The time taken to drill a 4-km well was 6 months (24 hours a day). This option seems to fit well in western Anatolia, where fracture and shear zones exist. In fact, being a tectonically dynamic terrain with several pull-apart basins, there are several shear zones and deep faults present in this region. The advantage, as described earlier, is the high-temperature regime at shallow depths. At about 2.5 km, the temperature of the granite is about 180°C higher than the UDDP site, and hence the final tariff of electricity will compete with the fossil fuel-based power and be lower than 6 Euro cents (see Table 3.4).

Option 4: (Figure 3.20)

Besides the above-explained cases, several other methods are being experimented with and discussed by several authors (Taleghani, 2013; Muir, 2020; Doran et al., 2021; Yuan et al., 2021; Budiono et al., 2022).

With maturing drilling technology, the option shown in Figure 3.20 will be the most advantageous in terms of cost, efficiency and low risk.

The electricity generated from the radiogenic granites can also be utilized to generate fresh water from seawater using the desalination method. Earlier work on similar granites from Saudi Arabia, Egypt, Eritrea, Djibouti and Tanzania indicates that desalination using geothermal energy is very efficient, cost-effective and the carbon footprint from this process is very low (Chandrasekhar et al., 2014; Chandrasekharam et al., 2015, 2016a, b, c, 2018a, b, 2019; Chandrasekharam, 2018, 2021a, b). The four common methods of desalination that are currently used worldwide are MED (multi-effect evaporation/distillation), MVC (mechanical vapour compression), MSF (multi-flash distillation) and reverse osmosis (RO) (Gude, 2015; Ghaffour et al., 2014). Currently, fossil fuel-based energy is being utilized for producing fresh water from the sea (Ghaffour et al., 2014; Fiorenza et al., 2003). Out of these four, MSF is the most cost-effective and efficient method with relatively low carbon emissions (Table 3.5). At present, around 12,500 desalination plants are in operation, generating 26 billion L/day of freshwater (Fiorenza et al., 2003; Chandrasekharam et al., 2018a, b).

The problem here is the cost of desalinated water, which varies with fuel costs. Using geothermal energy for desalination is recommended as the energy is free, perennial, cost-effective and has a very low-carbon footprint. The estimates show that 1 m^3 of desalinated water using fossil fuels (coal and oil) costs 1.4 euros, the cost is similar when geothermal energy is used as an energy input. But the advantage is that geothermal energy has a zero or very low-carbon footprint, unlike fossil fuels. Thus, there is no hidden cost in geothermal-based desalination. But in the case of fossil fuel-based desalination, cost is involved in mitigating environmental issues (Chandrasekharam et al., 2018b), like what the world is experiencing now.

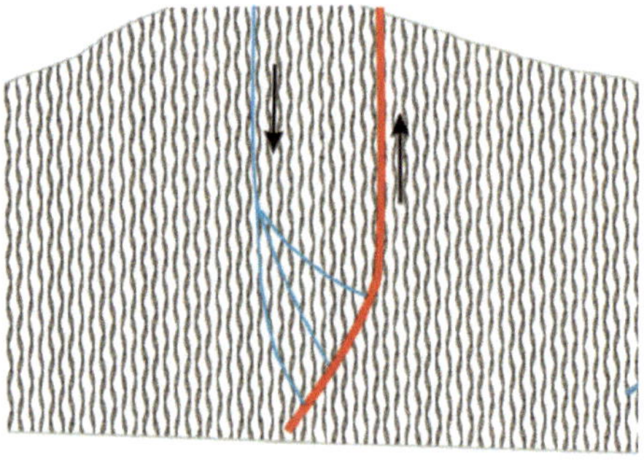

Figure 3.20 Schematic section shows the loop technology being developed to harness heat from the hot granites (for details, see the circled area in Figure 3.13). Adapted from Chandrasekharam et al. (2022).

Table 3.5 Energy consumption emissions and freshwater output of four common desalination methods

	MSF	*MED*	*MVC*	*RO*
Average capacity (10^3L/d)	25,000	10,000	3,000	6,000
Electrical energy (kWh/10^3L)	4	2	7	5
Production cost (Euro/10^3/L)	1.01	0.73	0.64	0.64
kg CO_2/10^3L (oil)	4	2	10	7

Source: Adapted from Chandrasekharam, 2021.

With maturing drilling technology and the availability of good EGS sites along western Anatolia, Turkey can adopt the sustainable development goals (SDGs) of the UN and can energy-water-food independent in the near future.

3.10 CONCLUSIONS

Turkey has huge amounts of renewable energy locked up in its high radiogenic granites spread over a cumulative area of 6,910 km^2 in western Anatolia. The country has already established a strong base for harnessing geothermal (hydrothermal) energy and can launch a strong geothermal programme to exploit its radiogenic granites. Due to the prevailing tectonic fabric and thermal regime, energy from the granites can be exploited at a much lower cost compared to the current fossil fuel-based energy cost. The current CO_2 emissions can be very well mitigated and aid in controlling global climate change. This will make the country energy independent and establish a strong economy. Coastal regions that are often affected by frequent droughts can overcome this problem by utilizing geothermal energy for the desalination of seawater. As shown above, desalinated water is cost-effective and can support a large population during water stress periods. The method of harnessing the heat from the granite has been described by various authors in this book. In conjunction with the field-based studies described here, numerical models related to the mechanism of the creation of fracture networks and the mechanism of extracting heat using supercritical CO_2 will open up avenues for investors to develop EGS-based power generation projects in western Anatolia.

ACKNOWLEDGEMENTS

This paper is part of the EGS project funded by TUBITAK (project No:120C079) through a fellowship grant to DC.

REFERENCES

Adam, A. 1987. Are there two types of the conductivity anomaly caused by fluid in the crust? *Phys. Earth Planet. Inter.* 45, 209–215.

Akal, C. 2013. Coeval Shoshonitic-ultrapotassic dyke emplacements within the Kestanbol Pluton, Ezine – Biga Peninsula (NW Anatolia). *Turkish J Earth Sci.* 22, 220–238.

Akay, E. 2008. Geology and petrology of the Simav Magmatic Complex (NW Anatolia) and its comparison with the Oligo-Miocene granitoids in NW Anatolia: Implications on tertiary tectonic evolution of the region, *Int. J. Earth Sci.* 98 (7), 1655–1675. https://doi.org/10.1007/s00531-008-0325-0.

Aldanmaz, E., Pearce, J.A., Thirlwall, M.F., and Mitchell, J.W. 2000. Petrogenetic evolution of late Cenozoic, post-collision volcanism in western Anatolia, *Turkey. J. Volcanol Geoher, Res.* 102, 67–95.

Altunkaynak, S., Sunal, G., Aldanmaz, E., Genç, C.S., Dilek, Y., and Furnes, H. 2012. Eocene granitic magmatism in NW Anatolia (Turkey) revisited: New implications from comparative zircon SHRIMP U–Pb and 40Ar–39Ar geochronology and isotope geochemistry on magma genesis and emplacement, *Lithos* 155, 289–309.

Angelier, J. 1990. Inversıon of field data in fault tectonics to obtain the regional stress. 3. A new rapid direct inversion method by analytical means, *Geophys. J. Int.* 103, 363–376. https://doi.org/10.1111/j.1365-246X.1990.tb01777.x.

Angelier, J. 1994. Fault slip analysis and paleostress reconstruction. In: P. Hancock (Ed.), *Continental Deformation*. Pergamon Press, Bristol, UK. pp. 53–100.

Avigrad, D., Jovilet, L., Garfunkel, Z., and Azanon, J.M. 1997. Back arc extension and denudation of Mediterranean eclogites, *Tectonics* 16, 924–941. https://doi.org/10.1029/97TC02003.

Aydin, I., Karat, H.I., and Koçak, A. 2019. Curie-point depth map of Turkey, *Geophys. J. Int.* 162 (2), 633–640. https://doi.org/10.1111/j.1365-246X.2005.02617.x.

Aydoğan, M.S., Coban, H., Bozcu, M., Akıncı, O. 2008. Geochemical and mantle-like isotopic (Nd, Sr) composition of the Baklan Granite from the Muratdag. Region (Banaz, Usak), western Turkey: Implications for the input of juvenile magmas in the source domains of western Anatolia Eocene. Miocene granites, *J Asian Earth Sci.* 33, 155.176.

Aysal, N., Öngen, S., Peytcheva, I., and Keskin, M. 2012. Origin and evolution of the Havran Unit, Western Sakarya basement (NW Turkey): new LA-ICP-MS U-Pb dating of the metasedimentary-metagranitic rocks and possible affiliation to Avalonian microcontinent, *Geodinamica Acta*, 25, 226–247, DOI: 10.1080/09853111.2014.882536

Baba, A., Bozkurt, C., Akkurt, G.G., Gudmarrson, T., Uzelli, T., Yakut, A.O., Döğerlioğlu, O. 2021. *Turkey–Assessment of Opportunities and Interest in Direct Uses of Geothermal Energy*, World Bank Report, The World Bank, New York, 132p.

Baba, A. and Chandrasekharam, D. 2022. Geothermal resources for sustainable development: Case study: Turkey, *Int. J. Energy Res.* 46, 1–18. https://doi.org/10.1002/er.7778.

Bağcı, M., Demirbilek, M., İlbeyli, N., Yıldız, A., and Kibici, Y. 2019. Geochronological and geochemical constraints and origin of the Tavşanlı Zone plutonic rocks (NW Turkey), *Turkish J. Earth Sci.* 28, 60–84. https://doi.org/10.3906/yer-1712-25.

Baumgartner, J., Lyman, M.P., and Andre, G. 1995. Drilling of hot and fractured granite at Soultz Sous Forets (France). *Proceedings, World Geothermal Congress, 1995*, Florence, Italy, Vol. 4.

Bingöl, E., Michel, D., and Gürol, A. 1982. Granitic intrusions in western Anatolia: A contribution to the geodynamic study of this area, *Eclogae Geol. Helv.* 75, 437–146.

Birkle, P. 1992. *Petrologie, geochimie und geochronologie des Miozanen magmatismus auf der Biga Halbinsel (Ezine-NW Turkei).* Doktora tezi. Tubingen Üniv. Unpublished PhD Thesis. 280 pp.

Black, K. N. 2012. Geochemical and Geochronological Relationships between Granitoid Plutons of the Biga Peninsula, NW Turkey. M.Sc. Thesis, the University of Texas at Austin.

Bozkurt, E. 2001. Neotectonics of Turkey-a synthesis, *Geodin. Acta* 14, 3–30. https://doi.org/10.1016/S0985-3111(01)01066-X.

Boztuğ, D, Harlavan, Y, Jonckheere, R, Can, I, and Sarı, R. 2009. Geochemistry and K-Ar cooling ages of the Ilıca, Cataldağ (Balıkesir) and Kozak (İzmir) granitoids, west Anatolia, Turkey. *Geol. J.* 44, 79–103.

Budiono, A., Suyitno, S., Rosyadi, I., Faishal, A., and Ilyas, A.X. 2022. A systematic review of the design and heat transfer performance of enhanced closed-loop geothermal systems, *Energies* 15, 742. https://doi.org/10.3390/en15030742.

Bozkurt, E. and Mittwede, S.K. 2005. Introduction: Evolution of continental extensional tectonics of western Turkey, *Geodin. Acta* 18 (3–4), 153–165. https://doi.org/10.3166/ga.18.153-165.

Bozkurt, E. and Oberhansli, R. 2001. Menderes Massif (Western Turkey): Structural, metamorphic and magmatic evolution-a synthesis, *Int. J. Earth Sci.* 89, 679–708.

Bozkurt, E., Winchester, J.A., Mittwede, S.K., and Ottley, C.J. 2006. Geochemistry and tectonic implications of leucogranites and tourmalines of the Southern Menderes Massif, Southwest Turkey, *Geodin. Acta* 19, 363–390.

Catlos, E.J., Baker, C.B., Sorensen, S.S., Jacob, L. and Çemen, I. 2011. Linking microcracks and mineral zoning of detachment-exhumed granites to their tectonomagmatic history: Evidence from the Salihli and Turgutlu plutons in western Turkey (Menderes Massif). *J. Struct. Geol.* 33, 951–969.

Celik, O.M., Gurer, O.F., and Ilknur, O.Z. 2009. Geochemical and petrographical properties of granitoidic rocks located between Geyve Gorge and Pamukova. *J. App. Earth Sci.* 8, 61–75.

Chandrasekhar, V., Omenda, P., and Chandrasekharam, D. 2014. High heat generating granites of Tanzania. *Proceedings 5th African Rift Geothermal Conference*, Arusha, Tanzania, October 2014, pp. 1–5.

Chandrasekharam, D. 2018. Water for the millions: Focus Saudi Arabia, *Water Energy Nexus* 1, 142–144.

Chandrasekharam, D. 2021a. Geothermal energy for food and water security for Yemen: A review, *Arab. J. Geosci.* 14, 253.

Chandrekharam, D. 2021b. Desalination of seawater using geothermal energy for food and water security: GCC and Sub-Sahara countries. *Proceedings of the World Geothermal Congress, 2020+1*, Iceland, June 2021, CD.

Chandrasekharam, D. and Baba, A. 2021. High heat generating granites of Kestanbol: Future Enhanced Geothermal System (EGS) province in Western Anatolia, *Turkish J. Earth Sci.* 30, 1033–1044. https://doi.org/10.3906/yer-2106-16.

Chandrasekharam, D. and Baba, A. 2022. Carbon dioxide emissions mitigation strategy through Enhanced Geothermal systems: western Anatolia, Turkey. *Environ. Earth Sci.* 81, 235–246. https://doi.org/10.1007/s12665-022-10345-5.

Chandrasekharam, D., Baba, A., Ayzit, A., and Singh, H.K. 2022. Geothermal potential of Kaymaz and Sivrihisar granites, Eskisehir region, western Anatolia, *Renew. Energy* 196, 870–882. https://doi.org/10.1016/j.renene.2022.07.035.

Chandrasekharam, D., Lashin, A., Al Arifi, N., and Al Bassam, A. 2016a. *Red Sea Geothermal Provinces*. CRC Press, London, UK. 220p.

Chandrasekharam, D., Lashin, A., Al Arifi, N., and Al-Bassam, A.M. 2018a. Energy and food security through desalination using geothermal energy: Eritrea, *Arabian J. Geosci.* 11, 523. https://doi.org/10.1007/s12517-018-3892-9.

Chandrasekharam, D., Lashin, A., Al Arifi, N., and Al-Bassam, A.M. 2018b. Desalination of seawater using geothermal energy for food and water security: Arab and Sub-Sahara countries. Chapter 4, 54p. In: G. Gnaneswar (Ed.), *Handbook on Sustainable Desalination Handbook – Process Design and Implementation Strategies*. Elsevier Pub., Butterwoth-Heinemann. 590p.

Chandrasekharam, D., Lashin, A., Al Arifi, N., and Al-Bassam, A.M. 2019. Securing food security of Djibouti through desalinated water using geothermal energy, *Energy Sustain. Soc.* 9, 24–35. https://doi.org/10.1186/s13705-019-0206-3.

Chandrasekharam, D., Lashin, A., Al Arifi, N., Al Bassam, A., El Alfy, M., Ranjith, P. G., Varun, C., and Singh, H.K. 2015. CO_2 emission and climate change mitigation using the enhanced geothermal system (EGS) based on the high radiogenic granites of the western Saudi Arabian shield, *J. African Earth Sci.* 112, 213–233.

Chandrasekharam, D., Lashin, A., Al Arifi, N., Al Bassam, A., and Varun, C. 2016b. Desalination of seawater using geothermal energy to meet future fresh water demand of Saudi Arabia, *Water Res. Manage.* 31, 781–792. https://doi.org/10.1007/s11269-016-1419-2.

Chandrasekharam, D., Lashin, A., Al Arifi, N., Al Bassam, A., Varun, C., and Singh, H.K. 2016c. Geothermal energy potential of eastern desert region, *Egypt Environ. Earth Sci.* 75, 697–709. https://doi.org/10.1007/s12665-016-5534-4.

Chandrasekharam, D. and Ranjith, P.G. 2020. CO_2 emissions from renewables: Solar pv, hydrothermal and EGS sources, *J. Geomech. Geophys. Geoenergy Georesour.* 6, 13–23. https://doi.org/10.1007/s40948-019-00135-y(0123456789().,-volV()0123458697().,-volV.

Cooper, G.T., Beardsmore, G.R., Waining, B.S., Pollington, N., and Driscoll, J.P. 2010. The relative costs of engineered geothermal system exploration and development in Australia. *Proceedings of the World Geothermal Congress*, Bali, Indonesia, April 2010, 17p.

Cotton, L., Doe, S., Gutmanis, J., Law, R., and Ledingham, P. 2018. An Overview of the United Downs Deep Geothermal Power (UDDGP) Project Authors. http://science4cleanenergy.eu/wp-content/uploads/2018/10/An-Overview-of-the-United-Downs-Deep-Geothermal-Power-UDDGP-Project.pdf.

Delaloye, M. and Bingöl, E. 2010. Granitoids from Western and Northwestern Anatolia: Geochemistry and modeling of geodynamic evolution, *Int. Geol. Rev.* 42 (3), 241–268. https://doi.org/10.1080/00206810009465081.

Demirbilek, M. 2012. Geochemical, Geochronological and Isotopic Constraints on the Intrusive Rocks in the Günyüzü, Sivrihisar and Kaymaz Regions (Eskişehir). PhD Thesis, Istanbul Technical University, Istanbul.

Demirbilek, M., Mutlu, H., Fallick, A.E., Sariiz, K., and Kibici, Y. 2018. Petrogenetic evolution of the Eocene granitoids in eastern part of the Tavşanlı Zone in northwestern Anatolia, Turkey, *Lithos* 314–315, 236–259. https://doi.org/10.1016/j.lithos.2018.06.003.

Diddglio, C., Guray, B.S., and Merdan, E. 2020. *Turkey Energy Outlook 2020.* International Centre for Energy and Climate, Sabanci University Pub., 452p. https://iicec.sabanciuniv.edu/.

Dilek, Y. and Altunkaynak, S. 2007. Cenozoic crustal evolution and mantle dynamics of post-collisional magmatism in Western Anatolia, *Int. Geol. Rev.* 49 (5), 431–453. https://doi.org/10.2747/0020–6814.49.5.431.

Dilek, Y., Altunkaynak, Ş., and Öner, Z. 2009. *Syn-extensional Granitoids in the Menderes Core Complex and the Late Cenozoic Extensional Tectonics of the Aegean Province.* Geological Society, London, Special Publications, 321, 197–223.

Dilek, Y., Altunkaynak, S., and Öner, Z. 2009. Syn-extensional granitoids in the Menderes core complex and the late Cenozoic extensional tectonics of the Aegean province, *Geol. Soc. Spec. Publ.* 321, 197–223. https://doi.org/10.1144/SP321.10.

Doran, H.R., Renaud, T., Falcone, G., Pan, L., and Verdin, P.G. 2021. Modelling an unconventional closed-loop deep borehole heat exchanger (DBHE): Sensitivity analysis on the Newberry volcanic setting, *Geotherm. Energy* 9, 4. https://doi.org/10.1186/s40517-021-00185-0.

Erdogan, B., Akay, E., and Ugur, M.S. 1996. Geology of the Yozgat region and evolution of the collisional Cankiri Basin, *Int. Geol. Rev.* 38, 788–806.

Erdogan, B., Akay, E., Hasözbek, A., Satır, M. and Siebel, W. 2013. Stratigraphy and tectonic evolution of the Kazdagı Massif (NW Anatolia) based on field studies and radiometric ages. *Int. Geol. Rev.* 55, 2060–2082, DOI: 10.1080/00206814.2013.818756.

Erdoğan, B. and Güngör, T. 2004. The problem of the core-cover boundary of the Menderes Massif and an emplacement mechanism for regionally extensive gneissic granites, western Anatolia (Turkey), *Turkish J. Earth Sci.* 13 (1), 15–36.

Erkul, F. and Erkul, T. 2010. Geology of the Early Miocene Alacamdad (Dursunbey-Balikesyr) magatic complex and implications for the western Anatolian extensional tectonic. *Mineral. Res. Expl. Bull.* 141, 1–25.

Erkül, S.T. and Erkül, F. 2012. Magma interaction processes in syn-extensional granitoids: The tertiary Menderes metamorphic core complex, western Turkey, *Lithos* 142, 16–33.

Esenli, F., Kadir, S., and Şansnli, B.E. 2019. Geochemistry of the Zeolite-rich Miocene pyroclastic rocks from the Gördes, Demirci and Şaphane Regions, West Anatolia, Turkey, 57, 1158–1172. https://doi.org/10.1134/S001670291911003X.

Fiorenza, G., Sharma, V.K., and Braccio, G. 2003. Techno-economic evaluation of a solar powered water desalination plant, *Energy Convers. Manag.* 44, 2217–2240.

Genç, S.C. 1998. Evolution of the Bayramiç magmatic complex, northwestern Anatolia. *J. Volcanol. Geother. Res.*, 85, 233–249.

Gessner, K., Gallardo, L.A., Wedin, F., and Sener, K. 2016. Crustal structure of the northern Menderes Massif, western Turkey, imaged by joint gravity and magnetic inversion. *Int. J. Earth Sci.* 105(7), 2133–2148.

Ghaffour, N., Lattemann, S., Missimer, T., Ng, K.C., Sinha, S., and Amy, G. 2014. Renewable energy driven innovative energy efficient desalination technologies, *Appl. Energy* 136, 1155–1165.

Giggenbach, W.F., Sano, J., and Wakita, H. 1993. Isotopic composition of He, and CO_2 and CH_4 contents in gases produced along the New Zealand part of a convergent plate boundaries, *Geochim. Cosmochim. Acta* 57, 3427–3456.

Goncuoglu, M.C., Koksal, S., and Floyd, P.A. 1997. Post-collisional A-type magmatism in the central Anatolian crystalline complex: Petrology of the Idis Dağı intrusives (Avanos, Turkey), *Turk J. Earth Sci.* 6 (2), 65–76.

Gorur, N., Oktay, F.Y., Seymen, I., and Sengor, A.M.C. 1984. Paleotectonic evolution of Tuzgölü basin complex, central Turkey. In: J.E. Dixon, A.H.F. Robertson (Eds.), *The Geological Evolution of the Eastern Mediterranean*. Geological Society Special Publications, London, Vol. 17, pp. 81–96.

Gücer, M.A., Arslan, M., Sherlock, S., and Heaman, L.M. 2016. Permo-Carboniferous granitoids with Jurassic high temperature metamorphism in Central Pontides, Northern Turkey, *Mineral. Petrol.* 110, 943–964. https://doi.org/10.1007/s00710-016-0443-5.

Gude, V.G. 2015. Energy storage for desalination processes powered by renewable energy and waste heat sources, *Appl. Energy* 137, 877–898.

Gulec, N. and Hilton, D.R. 2006. Helium and heat distribution in western Anatolia, Turkey: Relationship to active extension and volcanism, *Geol. Soc. Am.* 409, 1–16 (Special paper).

Gürbüz, A. and Gürer, O.F. 2009. Middle Pleistocene extinction process of pull-apart basins along the north Anatolian Fault zone, *Phys. Earth Planet. In.* 173, 177–180. https://doi.org/10.1016/j.pepi.2008.10.021.

Gurer, A., Gurer, F., Pince, A., and Ilkisik, M. 2001. Conductivity structure along the Gediz Graben, West Anatolia, Turkey: Tectonic implications, *Int. Geol. Rev.* 43, 1129–1144.

Gürer, A., Gurer, O., Pinçe, A., Ilkıs,ık, O.M., Gürer, O.F., Sangu, E., and Ozburan, M. 2006. Neotectonics of the SW Marmara gegion, NW Anatolia, Turkey, *Geol. Mag.* 143, 229–241. https://doi.org/10.1017/S0016756805001469.

Gurer, O.F., Kayrnak, N., Cakir, S., and Ozburan, M. 2003. Neotectonics of the southeast Marmara region, NW Anatolia, Turkey, *J. Asian Earth Sci.* 21, 1041–1051.

Gurer, O.F., Sangu, E., Gurer, A., and Akin, M. 2021. Late Cenozoic shift from extension to strike-slip stress regime in the west of the Biga Peninsula, NW Turkey, *J. Struct. Geol.* 148, 104–348.

Gurer, O., Sangu, E., Ozburan, M., Gurbuz, A., and Filoreau, N.S. 2013. Complex Basin evolution in the Gokova Gulf region: Implications on the Late Cenozoic tectonics of the southwest Turkey, *Int. J. Earth Sci (Geol Rundsch)*, 102, 2199–2221.

Gurer, O.F., Sangu, E., Ozburan, M., Gurbuz, A., Gurer, A., and Sinir, H. 2016. Plio-Quaternary kinematic development and paleo-stress pattern of the Edremit Basin, western Turkey, *Tectonophysics* 679, 199–210.

Hutter, G.W. 2020. Geothermal power generation in the world 2015–2020 update report. *Proceedings of the World Geothermal Congress*, Iceland, Reykjavik. pp. 1–17.

IEA. 2021a. Gas Market Report Q3-2021. International Energy Agency Report, IEA, France, Paris. 107p.

IEA. 2021b. Global Energy Review 2021: Assessing the Effects of Economic Recoveries on Global Energy Demand and CO_2 Emissions in 2021. International Energy Agency Report, IEA, France, Paris. 36p.

IEA. 2021c. *Turkey 2021: Energy Policy Review.* International Energy Agency, IEA, France, Paris. 191p.

Ilkisik, O.M. and Yenigün, H.M. 1997. Heat flow data and shallow thermal regime in western Turkey [abs.]. *Istanbul 97 International Geophysical Conference and Exposition (SEG, EAGE, Chamber of Geophysical Engineers of Turkey)*, Istanbul, Turkey. p. 74.

IPCC. 2018. *Global Warming of 1.5°C. An IPCC Special Report on the Impacts of Global Warming of 1.5°C Above Pre-Industrial Levels.* Intergovernmental Panel on Climate Change (IPCC), Geneva. https://www.ipcc.ch/sr15/.

Jackson, J. 1994. Active tectonics of the Aegean region, *Annu. Rev. Earth Planet. Sci.* 22, 239–271.

Jackson, J. and Mckenzie, J.P. 1988. The relationship between plate motions and seismic moment tensors, and the rates of active deformation in the Mediterranean and Middle East, *Geophys. J. Int.* 93, 45–73. https://doi.org/10.1111/j.1365-246X.1988.tb01387.x.

Karacık, Z., Yılmaz, Y., Pearce, J.A., and Ece, O.I. 2008. Petrochemistry of the south Marmara granitoids, northwest Anatolia, Turkey, *Int. J. Earth Sci. (Geol Rundsch)* 97, 1181–1200. https://doi.org/10.1007/s00531-007-0222-y.

Keles, S. and Bilgen, S. 2012. Renewable energy sources in Turkey for climate change mitigation and energy sustainability, *Renew. Sustain. Energy Rev.* 16, 5199–5206.

Kissel,C.andLaj,C.1988.TheTertiarygeodynamicalevolutionoftheAegeanarc:Apaleomagnetic reconstruction, *Tectonophysics* 146, 183–201. https://doi.org/10.1016/0040-1951(88)90090-X.

Koksal, S., Romer, R.L., Gonculu, M.C., Romer, R.L., and Koksal, F.T. 2004. Timing of post-collisional H-type to A-type granitic magmatism: U–Pb titanite ages from the Alpine central Anatolian granitoids (Turkey), *Int. J. Earth Sci. (Geol Rundsch)* 93, 974–989. https://doi.org/10.1007/s00531-004-0432-5.

Lachenbruch, A.H. 1968. Preliminary geothermal model of the Sierra Nevada, *J. Geophys. Res.* 73 (22), 6977–6989. https://doi.org/10.1029/JB073I022P06977.

Le Pichon, X. and Angelier, J. 1979. The Hellenic arc and trench system: A key to the neotectonic evolution of the eastern Mediterranean area, *Tectonophysics* 60, 1–42. https://doi.org/10.1016/0040-1951(79)90131-8.

McKenzie, D.P. 1972. Active tectonics of the Mediterranean regions, *Geophys. J. Roy. Astron. Soc.* 30, 109–185. https://doi.org/10.1111/j.1365-246X.1972.tb02351.x.

Meulenkamp, J.E., Wortel, M.J.R., Wamel, W.A.V., Spakman, W., and Strating, E.H. 1988. On the Hellenic subduction zone and the geodynamic evolution of Crete since the late Middle Miocene, *Tectonophysics* 146, 203–215. https://doi.org/10.1016/0040-1951(88)90091-1.

Muir, J.R. 2020. New opportunities and applications for closed-loop geothermal energy systems, *Geothermal Rising Bull.* 49 (4), 4–9.

O'Brien, P.J. 2019. Eclogites and other high-pressure rocks in the Himalaya: A review, In: P.J. Trelor, M.P. Searle (Eds.), *Himalayan Tectonics: A Modern Synthesis.* Geological Society Special Publications, London, Vol. 483, pp. 183–213.

O'Nions, R.K. and Oxburgh, E.R. 1988. Helium, volatile £uxes and development of continental crust, *Earth Planet. Sci. Lett.* 90, 331–347.

Ocakoglu, N., Demirbag, E., and Kuscu, I. 2004. Neotectonic structures in the area offshore of Alacati, Doganbey and Kusadasi (western Turkey): Evidence of strike-slip faulting in the Aegean extensional province, *Tectonophysics* 391, 67–83.

Okay, A. 2011. Tavsanli zone: The northern subducted margin of the Anatolide-Tauride block, *Miner. Res. Bull.* 142, 191–221.
Okay, A.I. and Tuysuz, O. 1999. Tethyan sutures of northern Turkey, In: B. Durand, B. Jolivet, L. Horvath, M. Seranne (Eds.), *The Mediterranean Basins: Tertiary Extension within the Alpine Orogen.* Geological Society Special Publications, London, Vol. 156, pp. 475–515.
Okay, A.I., and Satır, M. 2000. Coeval plutonism and metamorphism in the latest Oligocene metamorphic core complex in northwest Turkey. *Geol. Mag.* 137, 495–516.
Örgun, Y., Altınsoy, N., Gültekin, A.H., Karahan, G., and Çelebi, N. 2005. Natural radioactivity levels in granitic plutons and groundwaters in the Southeast part of Eskisehir, Turkey, *Appl. Radiat. Isot.* 63, 267–275. https://doi.org/10.1016/j.apradiso.2005.03.008.
Ozdamar, S. 2018. Petrology of Evciler Pluton (NW Turkey). *KSU J. Eng. Sci.* 2, 149–165 (in Turkish).
Özden, S., Över, S., Poyraz, S.A., Güneşm Y., and Pınar, A. 2018. Tectonic implications of the 2017 Ayvacık (Çanakkale) earthquakes, Biga Peninsula, NW Turkey, *Asian J. Earth Sci.* 154, 125–141.
Ozgenc, I. and Ilbeyli, N. 2008. Petrogenesis of the Late Cenozoic Eğrigöz Pluton in Western Anatolia, Turkey: Implications for magma genesis and crustal processes, *Int. Geol. Rev.* 50, 375–391. https://doi.org/10.2747/0020-6814.50.4.375.
Rebai, S., Philip, H., and Taboada, A. 1992. Modern tectonic stress field in the Mediterranean region: Evidence for variation in stress directions at different scales, *Geophys. Int.* 110, 106–140.
Ritchie, H. and Roser, M. 2020. *CO_2 and Greenhouse Gas Emissions.* Published online at Our World in Data.org. Retrieved from: https://ourworldindata.org/co2-and-other-greenhouse-gas-emissions (24 June 2021).
Rybach, L. 1976. Radioactive heat production: A physical property determined by the chemistry. In: R.G.I. Strens (Ed.), *The Physical and Chemistry of Minerals and Rocks.* Wiley-Interscience Publication, New York, pp. 245–276.
Şahin, S.Y., Örgün, Y., Güngör, Y., Göker, A.F., Gültekin, A.H., Karacik, Z. 2010. Mineral and Whole-rock Geochemistry of the KestanbolGranitoid (Ezine-Çanakkale) and its mafic microgranular enclavesin Northwestern Anatolia: Evidence of felsic and mafic magma interaction. *Turk. J. Earth Sci.* 19, 101–122.
Sangu, E., Gürer, O.F., and Gürer, A. 2019. Fault kinematic and plio-quaternary paleostress evolution of the Bakırcay basin, western Turkey, *Int. Geol. Rev.* 62, 1245–1261. https://doi.org /10.1080/00206814.2019.1642148.
Sangu, E., Gurer, O.F., and Gurer, A. 2020. Fault kinematic and plio-quaternary paleostress evolution of the Bakırçay basin, western Turkey, *Int. Geol. Rev.* 62, 1245–1261. https://doi.org /10.1080/00206814.2019.1642148.
Sasvári, A. and Baharev, A. 2014. SG2PS (structural geology to postscript converter) – A graphical solution for brittle structural data evaluation and paleostress calculation, *Computers Geosci.*, 66, 81–93.
Şengör, A.M.C., Tüysüz, O., İmren, C., Sakınç, M., Eyidoğan, H., Görür, N., Le Pichon, X., and Rangin, C. 2005. The North Anatolian fault: A new look, *Annu. Rev. Earth Planet. Sci.* 33, 37–112. https://doi.org/10.1146/annurev.earth.32.101802.120415.
Shin, T.A., Catlos, E.J., Jacob, L., and Black, K. 2013. Relationships between very high-pressure subduction complex assemblages and intrusive granitoids in the Tavşanlı Zone, Sivrihisar Massif, Central Anatolia, *Tectonophysics* 595, 183–197.
Speciale, P.A., Catlos, E.J., Yıldız, G.O., Shin, T.A., and Black, K.N. 2012. Zircon ages from the Beypazarı granitoid pluton (north central Turkey): Tectonic implications. *Geodinamica Acta*, 25, 162–182.
Somerville, M., Wyborn, D., Chopra, P., Rahman, S., Estrella, D., and Van der Meulen, T. 1994. *Hot Dry Rock Feasibility Study.* Energy Research and Development Corporation, unpublished report, Australia, Canberra. 214p.

Sunal, G. 2013. Devonian magmatism in the western Sakarya Zone, Karacabey region, NW Turkey, *Geodinamica Acta*, 25, 183–201.

Taleghani, A.D. 2013. An improved closed-loop heat extraction method from geothermal resources, *J. Energy Resour. Technol.* 135, 1–7.

Taymaz, T., Jackson, J.A., and McKenzie, D. 1991. Active tectonics of the north and central Aegean Sea, *Geophys. J. Int.* 106, 433–490. https://doi.org/10.1111/j.1365-246X.1991.tb03906.x.

Taymaz, T., Yılmaz, Y., and Dilek, Y. 2017. The geodynamics of the Aegean and Anatolia, *Geol. Soc. London Spec. Publ.* 291, 1–16. https://doi.org/10.1144/SP291.1.

Tezel, T., Shibutani, T., and Kaypak, B. 2013. Crustal thickness of Turkey determined by receiver function, *J. Asian Earth Sci.* 75, 36–45. https://doi.org/10.1016/j.jseaes.2013.06.016.

Topuz, G., Okay, A.I. 2017. Late Eocene - Early Oligocene two-mica granites in NW Turkey (the Uludağ Massif): Water-fluxed melting products of a mafic metagraywacke. Lithos, 268–271, 334–350.

Tuysuz, O., Dellaloglu, A.A., and Terzioglu, N. 1994. A magmatic belt within the Neo-Tethyan suture zone and its role in the tectonic evolution of northern Turkey, *Tectonophysics* 243, 173–191.

UNFCCC. 2015. https://unfccc.int/process-and-meetings/the-paris-agreement/the-paris-agreement/key-aspects-of-the-paris-agreement.

Uzel, B., Sozbilir, H., and Ozkaymak, C. 2012. Neotectonic evolution of an actively growing superimposed basin in western Anatolia: The inner bay of İzmir, Turkey, *Turk. J. Earth Sci.* 21, 439–471. https://doi.org/10.3906/yer-0910-11.

Yelden, E. and Voyvoda, E. 2015. *Low Carbon Development and Priorities for Turkey.* Sabancı University Istanbul Policy Centre, Report, Istanbul, Turkey. 69p.

Yılmaz, S., Örgün, Y., Güngör, Y., Göker, A. F., Gültekin, A. H., and Karacık Z. 2010. Mineral and whole-rock geochemistry of the Kestanbol Granitoid (Ezine-Çanakkale) and its mafic microgranular enclaves in Northwestern Anatolia: Evidence of felsic and mafic magma interaction. *Turkish J. Earth Sci.* 19, 101–122.

Yılmaz, Y. 2017. Morphotectonic development of Anatolia and the surrounding regions, *Am. Geophys. Union* 225, 11–91. https://doi.org/10.1002/9781118944998.ch2.

Yılmaz, Y., Genç, S. C., Gürer, O.F., Bozcu, M., Yılmaz, K., Karacık, Z., Altunkaynak, S., and Elmas, A. 2000. When did the western Anatolian grabens begin to develop? A. Tectonics and magmatism in Turkey and the surrounding area, *Geol. Soc.* 173, 353–384. https://doi.org/10.1144/GSL.SP.2000.173.01.17.

Yuan, W., Chen, Z., Grasby, S., and Little, E. 2021. Closed-loop geothermal energy recovery from deep high enthalpy systems, *Renew. Energy* 177, 976–991.

Chapter 4

Introduction to conceptual studies on hot dry rock geothermal energy extraction aided by prospective tunnelling

K. Ji, Y. Tao, F.J. Huang, Y. Liu, and H. Li
Dalian University of Technology

CONTENTS

4.1 INTRODUCTION

Fossil energy has long been a "double-edged sword." While driving the world's rapid economic development, it has also made the world's energy crisis and environmental contamination increasingly serious. Therefore, the search for renewable and clean energy that can replace fossil fuels has become an important task for social development [1].

Compared with other renewable energies, such as solar and wind energy, geothermal energy can be independent of seasonal, climatic, diurnal and other factors and has huge reserves, enabling a stable and uninterrupted energy supply [2]. Considering the many advantages of geothermal energy, it has received widespread attention from countries around the world. Geothermal resources can be divided into three categories according to their depth in the earth's strata: shallow geothermal energy, hydrothermal resource and hot dry rocks (HDRs) [3]. In particular, due to its huge resource reserves, HDR is expected to play a more important role in geothermal power generation or district heating than the other two types of geothermal energy. Conservative estimates put the amount of geothermal energy stored in HDR at 1.4×10^{25} J in the USA and 2.52×10^{25} J in China [4]. However, HDR is characterized by its difficulty of development, and current HDR development is still in the experimental stage. Consequently,

DOI: 10.1201/9781003271475-4

how to achieve large-scale exploitation of HDR will be the promising direction of future geothermal development research [5].

In 1970, the concept of enhanced geothermal systems (EGSs) was first introduced to extract heat from HDR by the Las Amos National Laboratory in the USA [6]. There have been several typical cases of EGS over the last 40 years, such as Fenton Hill, Hijiori, Ogachi, Cooper Basin and Soultz. In particular, the most successful EGS case is located at the Soultz EGS site, which proved that the idea of using EGS technology to exploit geothermal energy was sound, but its power generation was only 1.5 MW [7]. “If the future method of geothermal energy exploitation is EGS, then what is the future of EGS?”, Rybach, former President of the International Geothermal Association, asked this question in his keynote presentation at the World Geothermal Congress in Bali, Indonesia. The question expressed deep concern about the bottlenecks in the development of EGS and the persistent failure to achieve large-scale exploitation of HDR. In conclusion, it is difficult to realize the large-scale exploitation of the HDR only by the EGS, and it can be seen that innovative breakthroughs are urgently needed in the research of the deep geothermal exploitation scheme.

In this chapter, the existing EGS technology is reviewed, including its principles, technical characteristics and shortcomings. Then, aiming at a solution to the bottleneck of EGS, we proposed a novel geothermal energy extraction scheme turning petroleum into mining technology, i.e., excavation-based EGS (EGS-E), to help realize the large-scale exploitation of HDR. In this chapter, drillhole or well-based EGS exploitation technology is called EGS-D in order to differentiate it from excavation-based EGS (EGS-E) exploitation that is proposed in this article.

4.2 EXISTING DRILLING-BASED EGS

4.2.1 Brief introduction

HDR is distributed in the underground within a depth range of 3–10 km, where the formation temperature reaches 150°C–650°C. Because of the extremely low natural permeability of HDR, geothermal energy cannot be economically extracted, so the development of HDR requires the establishment of EGS. Generally, EGS-D typically consists of injection wells, production wells and heat reservoirs created by artificial enhancement techniques, as shown in Figure 4.1 [8].

Hydraulic fracturing was used as the primary artificial enhancement technique to create fracture networks in the HDR to connect these wells and create more heat transfer area. Cold working fluid is injected into the artificial heat reservoir through the injection wells. Then, the working fluid permeates through the fracture networks and exchanges heat with the high-temperature rocks to extract the energy, returning as hot fluid through the production wells.

4.2.2 Development status of EGS-D

According to statistics [9–11], 43 HDR development projects have been implemented worldwide. These projects are mainly located in 14 countries in Europe, North America, Australia and Asia, with Germany and the USA being the countries with the most implemented HDR development projects. The number of projects implemented in each country is shown in Table 4.1.

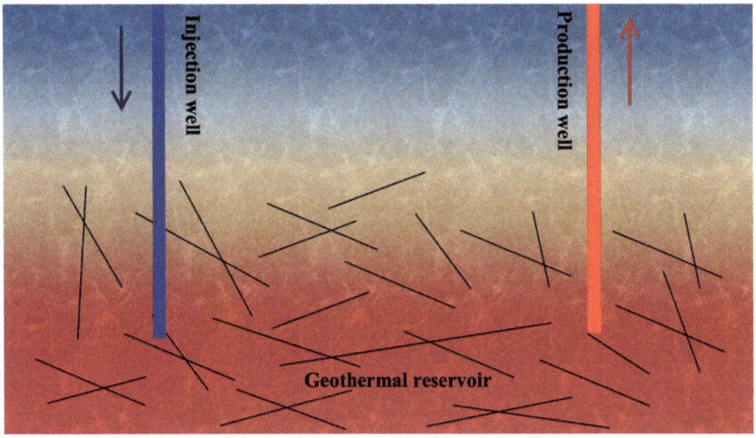

Figure 4.1 Schematic diagram of the EGS-D.

Table 4.1 Statistics of 43 global HDR development projects (ranked by startup time)

No.	*Country*	*Project name*	*Status*	*Year of startup*	*Generating capacity (MW)*
1	Italy	Lardarello	In operation	1970s	700 (includes hydrothermal power generation)
2	United States	Fenton Hill	Terminate	1974–1992	0.06
3	Germany	Falkenberg	Terminate (low output temperature)	1977–1986	
4	Germany	Bad Urach	Terminate (lack of funding)	1977–2008	
5	France	Le Mayet	Terminate (temperature)	1978–1986	
6	Germany	Bruchsal	In operation	1983	0.55
7	Germany	Neustadt-Glewe	In operation	1984	0.21
8	France	Soultz-sou-Forêts	In operation	1984	1.5
9	Sweden	Fjällbacka	Terminate (low output temperature)	1984–1995	
10	United Kingdom	Rosemanowes	Terminate (low output temperature)	1984–1992	
11	Japan	Hijiori	Terminate (water loss)	1985–2002	0.13
12	Austria	Altheim	In operation	1989	1
13	French West Indies	Bouillante	In operation	1996	15 (includes hydrothermal power generation)

(*Continued*)

Table 4.1 (Continued) Statistics of 43 global HDR development projects (ranked by startup time)

No.	*Country*	*Project name*	*Status*	*Year of startup*	*Generating capacity (MW)*
14	Australia	Hunter valley	Terminate (lack of policy and financial support)	1999–2015	
15	Germany	Groβ-Schnebeck	In operation	2000	1
16	Salvador	Berlín	In operation	2001	Plan to 6
17	United States	Coso	Terminate (drilling and fracturing accidents)	2002–2012	
18	Germany	GeneSys Horstberg	Terminate	2003	
19	Germany	GeneSys Hannover	Only for experimental research	2003–2007, restarted in 2015	
20	Germany	Landau	Pause (earthquake in 2013)	2003–2013	2.9–3
21	Australia	Cooper Basin Habanero	Terminate (lack of funding)	2003–2013	1
22	Germany	Unterhaching	In operation	2004	Plan to 3.4
23	Switzerland	Basel	Terminate (accompanying the earthquake)	2005–2009	
24	Australia	Paralana	Pause (lack of funding)	2005–2014	
25	Germany	Insheim	In operation	2007	Plan to 4.8
26	United States	Southeast Geysers	Terminate (wellbore collapse)	2008–2009	
27	United States	Bradys	Terminate	2008–2015	
28	Switzerland	St. Gallen	Terminate (accompanying the earthquake)	2009–2014	
29	United States	Raftriver	In operation	2009	Plan to 5
30	United States	Northwest Geysers	In operation	2009	Plan to 3.5
31	United Kingdom	Redruth	In operation	2009	Plan to 10
32	United Kingdom	Eden	Under planning	2010	Plan to 4
33	United States	Newberry Volcano	In operation	2010	Plan to 35
34	Korea	Pohang	Terminate (earthquake)	2010–2017	
35	China	Gonghe	In experiment (drilled)	2011	
36	United States	Milfod	In experiment (not yet drilled)	2011	

(Continued)

Table 4.1 (Continued) Statistics of 43 global HDR development projects (ranked by startup time)

No.	*Country*	*Project name*	*Status*	*Year of startup*	*Generating capacity (MW)*
37	Germany	Mauerstetten	Only for experimental research (the effects of hydraulic fracturing are bad, lack of funding)	2011–2012, restarted in 2015	
38	The Franco-German cooperation	GEOSTRAS	Under planning	2012	Plan to 6.7
39	United States	Desert Peak	Terminate	2012–2013	1.7
40	United States	Milford	In experiment (not yet drilled)	2015	
41	Hungary	Szeged	Under planning	2016	Plan to 8.9
42	China	Matouying	In operation	2019	0.2 (thermal power)
43	Canada	Eavor-Lite	In operation (commercial demonstration facility)	2019	

The earliest of the 43 projects is the Fenton Hill project in the USA, which carried out trials of EGS-D in 1973. During the first phase of the Fenton Hill project (1973–1979), the maximum borehole depth was 2,600 m, the reservoir temperature was 185°C and the production well was capable of producing high pressure fluids at 135°C–140°C with flow rates ranging from 7 to 16 kg/s, allowing a 60 kw duplex power station to be established. The project has played a huge leading role in the exploitation of HDR using EGS-D [12]. The 1980s saw the first surge in HDR development projects, with a total of eight projects being initiated. The Soultz EGS-D is the most successful HDR development in the world to date, with an installed capacity of 1.5 MW of stable commercial-scale power generation. EGS-D developed at a slower pace in the 1990s, with one project in the French West Indies and one in Australia. The 21st century has seen a second boom in HDR exploration and exploitation, with 30 HDR development projects initiated worldwide, mainly in the USA, Germany and Australia. Although China is rich in HDR resources, the research has started late. Since the first research project on high-temperature drilling technology for HDR was launched in 2010, China has conducted HDR drilling in the Gonghe Basin, Zhangzhou, Tangshan, Haikou, etc., and explored a number of high-quality HDR geothermal fields. Among these, the Gonghe Basin has the highest geothermal resource temperature, with an HDR of 236°C drilled at 3,705 m [1]. Additionally, the Ma Touying EGS project will achieve China's first experimental power generation by HDR geothermal energy in 2019. The M-1 well of the Ma Touying EGS project is at a depth of 4,000 m and has a bottom hole temperature of 151.25°C. The researchers carried out the experiments of heat extraction by an ultra-long gravity heat pipe and temperature difference power generation in the M-1 well. Under the condition that the average temperature of the rock in the underground heat extraction section is 119℃, saturated water vapour with a maximum temperature of 90℃ was obtained, and the continuous heat extraction power is close to 200 KW [13].

A breakdown of the current status of these HDR developments shows that only 16 projects are currently in operation. The main reasons are as follows: Firstly, some projects were terminated due to lack of continuous funding, such as the Fenton Hill project in the USA and the Cooper Basin in Australia. Secondly, some projects were terminated due to microseismic problems induced by hydraulic fracturing or environmental problems, such as the Pohang project in Korea and the Basel project in Switzerland. Thirdly, some of the projects were purely experimental research projects that were terminated after their objectives were achieved, such as the Rosemanowes project in the UK [14].

4.2.3 Shortcomings of EGS-D

In general, the development of HDR is entering a boom period globally, but the large number of terminated projects mentioned above also shows that there are many technical challenges associated with HDR development. In particular, these challenges stem from the inability to achieve large-scale development and poor controllability of current EGS-D technology.

The volume of geothermal reservoirs is small and difficult to control. The current system uses mainly boreholes and hydraulic fracturing techniques to construct geothermal reservoirs, so the number of injection and production wells and the fracturing distance between them greatly limit the volume of geothermal reservoirs. The longest fracturing distance is currently only 100m, resulting in insufficient effective stimulation volume being generated. Therefore, it is more difficult to rely on increasing well spacing to expand the volume of geothermal reservoirs. Another way of increasing the volume of geothermal reservoirs is to increase the number of wells. In order to develop a relatively large-scale EGS, say up to 200 MW, it may require around 100 pairs of deep wells. From the point of view of current drilling technology, a large-scale EGS-D development is not economically feasible [15].

Low production rate of working fluid. The working fluid returns to the surface from the production wells after exchanging heat with geothermal reservoirs. The small diameter of wells means that the volume of working fluid injected and recovered is extremely limited. According to Jung's research, the conditions that need to be met for EGS to be able to produce electricity commercially are that the working fluid flow rate should be 50–100 L/s and the power should be guaranteed for 3–10 MW for more than 25 years [16]. EGS-D typically has a borehole diameter of only about 10cm and a production flow rate of only 20 L/s, which greatly hampers the efficiency of heat extraction [10].

Poor controllability of the effects of hydraulic fracturing. The distribution and extension of fractures in geothermal reservoirs are difficult to control due to the different physical properties of the rocks, differences in natural fractures and various uncertainties in the fracturing process [10]. Once fractures created by hydraulic fracturing are connected to a large fault already in the formation, it can cause working fluid injected into the reservoir to flow out of the reservoir and not be produced from the producing well, which is known as water loss. A similar problem was encountered at Rosemanowes in the UK, where attempts to improve geothermal reservoir permeability by increasing injection pressure resulted in the growth of the fracture system far away from the geothermal reservoir, increasing water loss without improving connectivity. During the operation of the system, the injected working fluid will always flow quickly along the cracked channel with a large opening and good connectivity and then be discharged from the production wells, causing short-circuit circulation. Short-circuit circulation is one of the main issues affecting the operational effectiveness and economic viability of EGS-D, and it will cause a large part of the rock fracture networks to fail to perform the heat exchange function [17].

Uncontrollable safety and contamination issues. In order to improve the fracturing effects, chemicals are sometimes added to the fracturing fluid, which can cause serious contamination of groundwater. Fracturing activity will disturb the in situ stress, which may trigger seismic activity. In Basel, Switzerland, 3,500 microseismic events occurred between 2006 and 2007 as a result of the EGS's development. In 2007, an earthquake of magnitude 5.4, the highest magnitude estimated to be related to EGS-D, struck the northern part of Pohang city and suspended the local geothermal power generation project, causing a direct economic loss of more than $47 million [18].

Uncontrollable income risk. Warren Buffett once said that direction is more important than speed. To give a simple example, if you invest $100, you will make a profit of $5 in a year. Although the rate of return is only 5%, it is a steady and profitable return. Furthermore, when we expand the scale of investment to $10 billion, the income can reach $500 million, which will be a considerable income. As we are all familiar, for large investment projects, a stable rate of return is the most critical parameter to measure the prospects of the project. Because its scale is very large and its operating cycle is very long, we can obtain high returns over a long period of time by increasing its scale. However, the current EGS-D suffers from water losses, short-circuit circulation, high randomness in fracture distribution and seismic-induced shutdowns, all of which can lead to an inability to evaluate yields more accurately. In a sense, current HDR development by EGS-D is more of a gamble, and as such, it is difficult to gain the favour of commercial investors, which seriously affects the commercialization and application of EGS.

4.3 THE INTRODUCTION OF EGS-E

From the above shortcomings, we may find the future development direction of EGS, i.e., large-scale and small randomness. Therefore, the existing EGS of using drilling and hydraulic fracturing will no longer be applicable. To this end, we have proposed a method to try to provide a new solution for the development of HDR from the perspective of using mining methods.

4.3.1 Core principles

For almost all new technologies of HDR exploitation, researchers are most concerned about how to build geothermal reservoirs and how to achieve efficient heat transfer. This is because these two aspects are closely related to the effectiveness of heat extraction. Traditional methods of heat extraction use drilling and hydraulic fracturing to build geothermal reservoirs, but the EGS-E has largely moved away from reliance on them. For the EGS-E, it uses mining technology to form tunnel systems with specific structures in intact HDR, thus completing the construction of geothermal reservoirs that can be further exploited. The EGS-D uses a direct "fluid-rock" heat exchange system, while the EGS-E uses a two-stage heat exchange system of "fluid-rock (tunnel heat exchange subsystem) and fluid-fluid (pipe heat exchange subsystem)."

4.3.2 Physical components of EGS-E geothermal reservoir

Figure 4.2 shows the basic unit of geothermal energy extraction of EGS-E. The physical structure of the EGS-E geothermal reservoir consists of a shaft system, a tunnel system, a piping system, a drive system and other supporting structures (Figures 4.3 and 4.4).

In the above diagram:

1: Shaft; 2: Connection tunnel; 31: Maintenance tunnel; 32: Heat exchange tunnel; 33: Near-field heat extraction tunnel; 34: Far-field heat extraction tunnel; 41: Hole; 42: Caved rubble area; 51: Gate of Maintenance tunnel; 52: Gate of heat extraction tunnel and heat exchange tunnel; 61: 1# Injection pipe; 62: Pump; 63: 2# Injection pipe; 64: Production pipe; 65: Heat exchange pipe; 7: Lifting equipment; 8: Thermal energy utilization equipment.

The shaft system consists of a large diameter shaft and lifting device. The large diameter shaft is excavated from the surface to the target depth and is the main access point for lifting, transportation, ventilation and communication with the surface. The

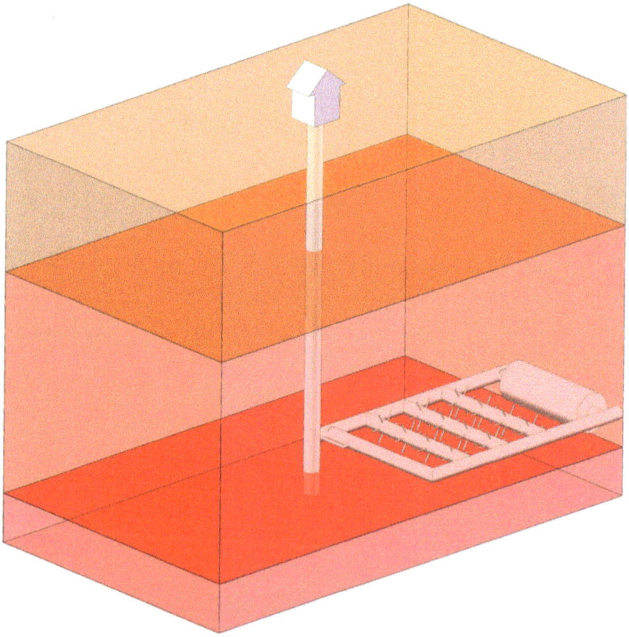

Figure 4.2 Three-dimensional schematic diagram of our concept.

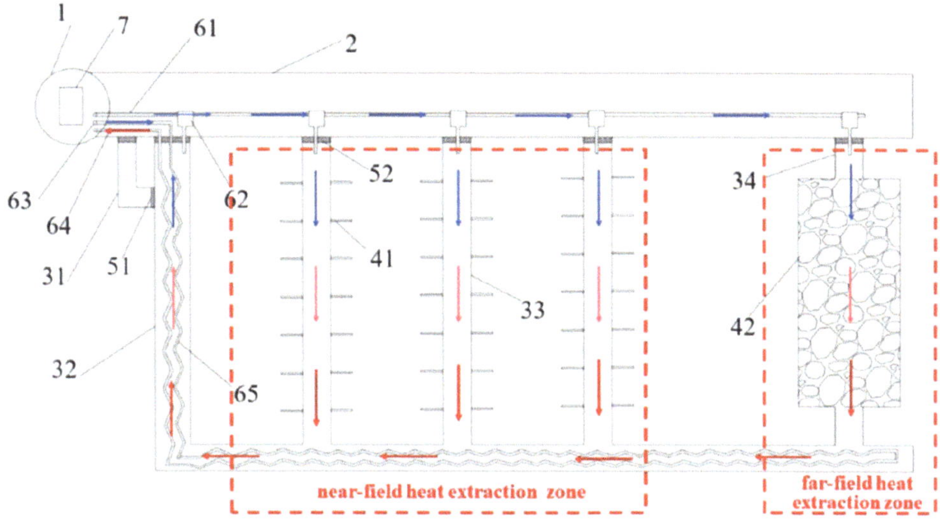

Figure 4.3 Two-dimensional plane diagram of our concept.

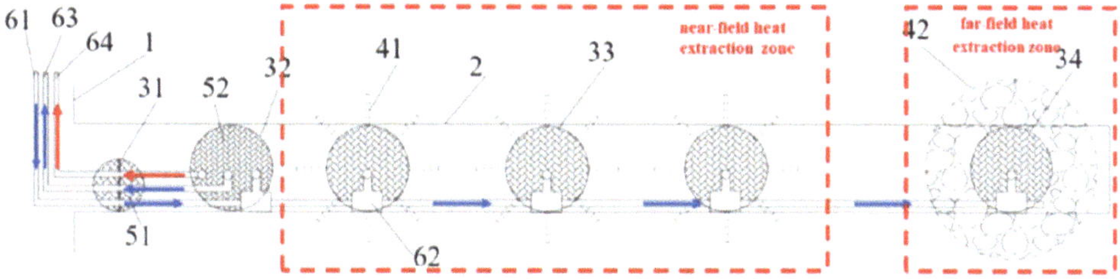

Figure 4.4 Two-dimensional profile diagram of our concept.

tunnel system consists of connection tunnels, heat extraction tunnels, heat exchange tunnels and maintenance tunnels. The connection shaft is the main shaft of the EGS-E and functions as a framework for the horizontal stratification of the geothermal reservoirs. Multiple layers of connection tunnels can be arranged longitudinally along the shaft as required. The heat extraction tunnel is the core of the entire tunnel system and is divided into the near-field heat extraction tunnels and the far-field heat extraction tunnels. As the name suggests, it is the main space for the heat exchange between fluid and surrounding rock (first stage of heat exchange). In order to increase the heat exchange area, different enhancement methods are adopted for the near-field heat extraction tunnels and the far-field heat extraction tunnels. Drilling holes in the wall of the near-field heat extraction tunnels to increase the heat contact area. In the far-field heat extraction tunnels, blasting is carried out in an orderly manner, causing the rocks around the tunnel to collapse and break to form large-volume gravel to increase the heat contact area. Where there are multiple tunnel system levels in geothermal reservoirs, the locations of the heat extraction tunnels at the upper and lower levels are staggered to avoid mutual interference. The function of the heat exchange tunnel is to gather the high-temperature fluid flowing out of the heat extraction tunnels and provide a place for the second-stage heat exchange of "fluid-fluid." The heat exchange tunnel is the intersection of a two-stage heat exchange system. It can belong to the tunnel heat exchange system or the pipe heat exchange system. The maintenance tunnel connects the connection tunnel with the heat extraction and heat exchange tunnels to facilitate access for maintenance. Opening gates at the entrances and exits of said maintenance tunnel, which are normally closed when the system is in operation and open when used for maintenance and dredging work. The entrance gates to the rest of tunnels are provided with multiple holes for various types of pipes to pass through, facilitating connections to the outside. Piping system refers to all kinds of pipelines located in shafts, connection tunnels, and heat exchange tunnels, mainly including 1# injection pipelines, 2# injection pipelines, collection pipelines and heat exchange pipelines. The drive system mainly includes fluid pumps and other equipment, and its function is to promote the working fluid to circulate in the heat extraction and heat exchange tunnels.

According to needs, the basic unit of geothermal energy extraction can be arranged around the shaft and the connecting tunnel in any combination so as to form a larger-scale geothermal reservoir (see Figures 4.5, 4.6 and 4.7).

4.3.3 The process of heat extraction

The system operation process is a reflection of the heat exchange process. The heat exchange system of EGS-E consists of a tunnel heat extraction subsystem and a pipe heat exchange subsystem. The extraction of deep geothermal energy is accomplished

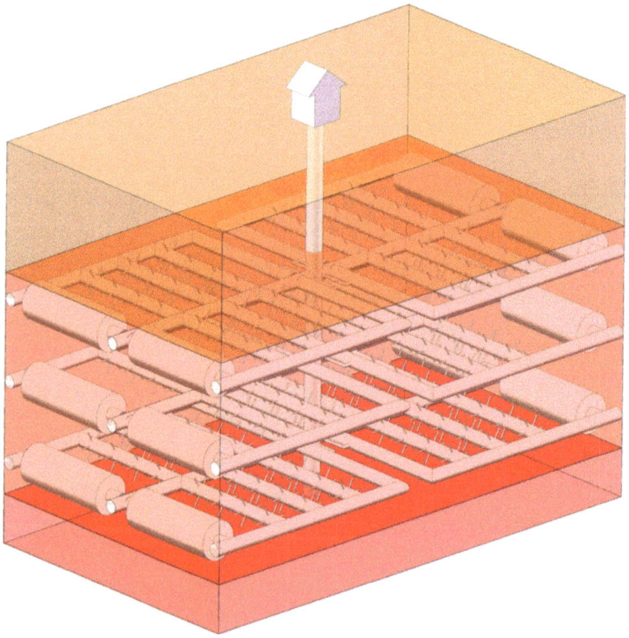

Figure 4.5 Combination of the basic units of geothermal energy extraction.

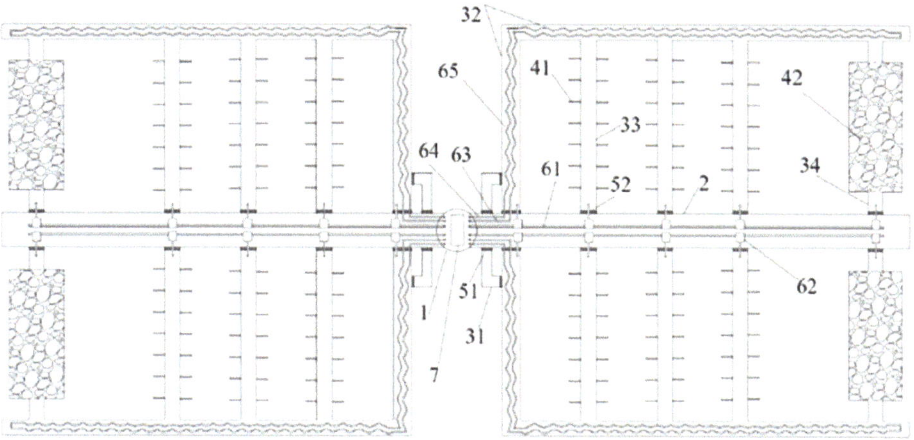

Figure 4.6 Two-dimensional plane diagram of combination.

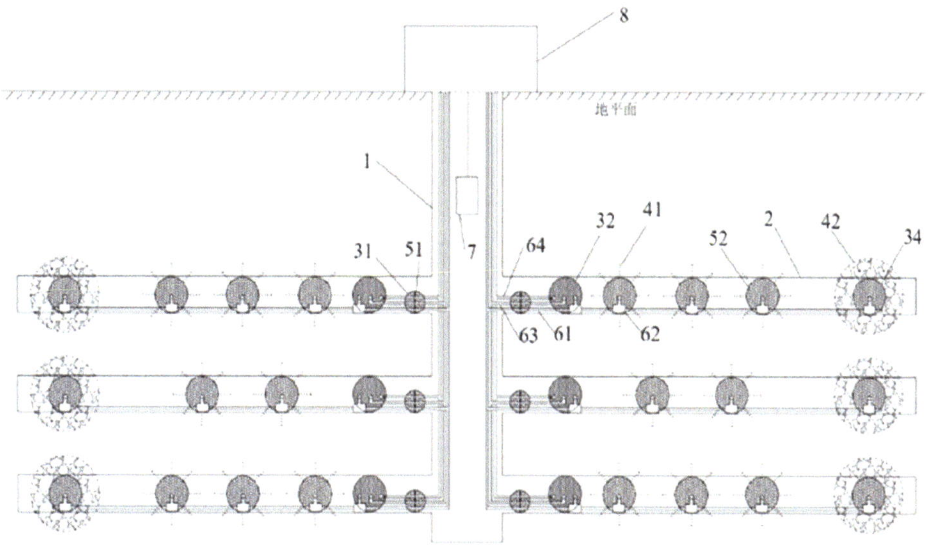

Figure 4.7 Two-dimensional profile diagram of combination.

through a two-stage heat exchange: "fluid-rocks" (tunnel heat extraction subsystem) and "fluid-fluid" (pipe heat exchange subsystem).

The tunnel heat exchange subsystem is the first-stage heat exchange system of EGS-E, which is mainly composed of near- and far-field heat extraction tunnels, heat exchange tunnels, driving devices, 1# injection pipes, and 1# working fluid filled in the above tunnels. The 1# working fluid is transported through the 1# injection pipeline to heat extraction and heat exchange tunnels; when the 1# working fluid fills the tunnels, the 1# working fluid injection is stopped. The drive devices outside the gates are opened to drive the 1# working fluid to start circulating in the above tunnels. The 1# working fluid conducts sufficient heat exchange with the tunnel wall, borehole wall and gravel in near- and far-field heat extraction tunnels. After full heat exchange, the 1# working fluid converges into the heat exchange tunnel, thus completing the tunnel heat exchange.

The pipe heat exchange subsystem is the second-stage heat exchange system of EGS-E, which is mainly composed of injection and collection pipes arranged in the shaft and connecting tunnel and spiral heat exchange pipes arranged in the heat exchange tunnel. The 2# working fluid is injected into the heat exchange pipe through the 2# injection pipeline, and the 2# working fluid in the heat exchange pipe exchanges heat with the 1# working fluid filled in the heat exchange tunnel and vaporizes. Due to the thermosiphon effect (see Figure 4.8), the vaporized 2# working fluid floats to the ground through the collection pipe under the action of thermal buoyancy on the heat utilization equipment. After being used, the 2# working fluid becomes cold and liquefied and flows into the heat exchange pipe through the 2# injection pipe again under the action of gravity to perform self-circulation and reciprocate in this way to realize the pipe heat exchange.

In summary, during the geothermal energy extraction process of EGS-E, the heat is first transferred to the 1# working fluid through the surface of tunnels, boreholes and gravel. Secondly, the 1# working fluid acts as a carrier medium to collect heat for the heat exchange tunnel. Finally, the heat is transferred to the 2# working fluid in the heat exchange pipeline and then transferred to the utilization equipment located on the ground (Figure 4.9).

4.3.4 Assessment of technical advantages

Customizable and large-scale volume of geothermal reservoirs. The volume of geothermal reservoirs in EGS-D is affected by a combination of factors such as hydraulic fracturing and the geological environment, making it difficult to form geothermal reservoirs on a large scale and difficult to accurately control the volume of geothermal

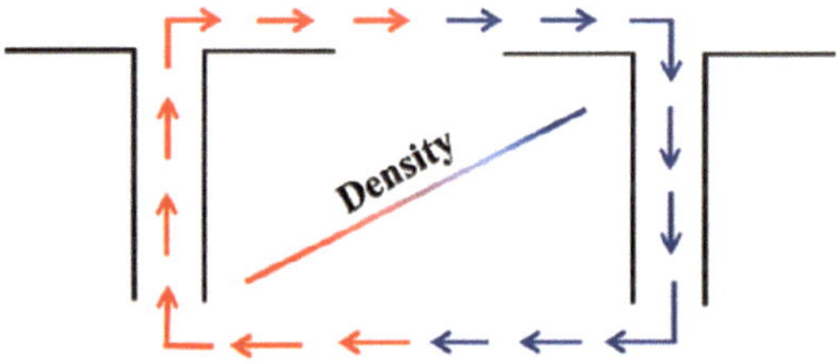

Figure 4.8 Diagram of the thermal siphon effect.

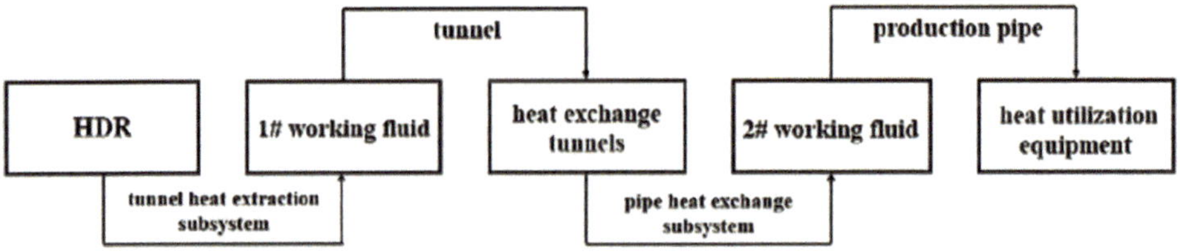

Figure 4.9 Flowchart of heat transfer in EGS-E.

reservoirs. The Fenton Hill EGS-D in the USA has an estimated geothermal reservoir volume of 0.035 km^3, far short of the minimum geothermal reservoir volume required for commercial development [19]. The EGS-E uses a tunnel system to delineate the geothermal reservoir. By adjusting the structural parameters of the tunnel system, such as length, diameter, quantity, etc., to establish different volumes of geothermal reservoirs more precisely to meet different requirements. The longer the length of the tunnel, the higher the extraction level, the larger the scale of the delineated geothermal reservoirs, and vice versa.

On the other hand, the EGS-E structure is highly expandable and can be expanded by excavating new far-field heat extraction tunnels to the distal end of the structure depending on the decay of the geothermal reservoirs temperature, thus increasing the volume of the geothermal reservoirs.

Stable flow rate of working fluid. In terms of flow stability, during the operation of EGS-D, the obvious water loss phenomenon will make the total amount of working fluid produced insufficient and cause large fluctuations. For EGS-D projects to date, water loss has exceeded 10% in all site tests except Soultz, and over 70% water loss has been observed in the Hijiori EGS-D in Japan. For the EGS-E, due to the use of two-stage heat exchange, the fluid flow stability includes the flow stability of the 1# fluid in the tunnel and that of the 2# working fluid in the heat exchange pipeline. Due to the convenience of tunnel construction, certain engineering methods can be adopted to reduce the loss of 1# working fluid. Spraying low-permeability materials on the wall surface of tunnels in the near-field heat extraction zone can prevent the loss of 1# working fluid. In the far-field heat extraction zone, although low-permeability materials cannot be used to seal the wall, detection methods can be used to keep the construction away from the original fissure passages to reduce fluid loss as much as possible. For the 2# working fluid, as it flows within a closed pipe, it is difficult to experience fluid loss unless there is a break in the pipe. In summary, the EGS-E method has a stable fluid flow rate.

Further, thanks to the use of large diameter tunnels and pipes as fluid flow paths, the limitations of small diameter boreholes in traditional methods are broken down, and a high flow rate and steady circulation of working fluid can also be achieved.

High-temperature working fluid. The working fluid temperature depends primarily on the heat absorbed by the working fluid during the heat exchange process, i.e., on the effective heat exchange area of the system and how the system is operated. EGS-E uses the wall surface of the tunnel and the borehole for heat transfer in the near-field heat extraction zone, so the structural parameters of the tunnel and the borehole can be adjusted to provide sufficient heat exchange area in the near-field heat transfer zone. In the far-field heat extraction zone, working fluids are mainly exchanged with the new wall surface and gravel created by the blasting of the tunnel, so the blasting parameters can be adjusted to control the size of the grave and the impact of the blasting to ensure that the far-field heat extraction zone has sufficient heat exchange area. Since both the near- and far-field heat exchange zones can ensure sufficient heat exchange area, it can ensure that the 1# working fluid circulating in the tunnel has a high temperature. Further, the long spiral heat transfer pipe located in the heat transfer tunnel ensures that the second-stage heat exchange of "fluid-fluid" has sufficient heat exchange area to ensure that the working fluid returned to the ground utilization equipment is of high temperature.

On the other hand, EGS-E can also achieve stable and continuous production of high-temperature working fluid by adjusting the operation mode. Thanks to the distributed layout of the structure, the rotation system can be used for exploitation to avoid the premature occurrence of thermal breakthroughs.

Adjustable heat extraction rate. Circulation rate of the working fluid in the system plays an important role in the rate of heat extraction. EGS-D can't even guarantee a stable output flow rate, so the controllable circulation speed is more difficult to achieve, and thus it is impossible to realize the adjustment of the heat extraction rate. For EGS-E, the heat extraction rate can be adjusted in a controlled manner according to changes in external demand. The heat extraction rate is regulated by adjusting the driving device to control the circulation rate of 1# working fluid or by adjusting the 2# working fluid injection rate.

Less energy consumption. For EGS-D, the working fluid has to overcome huge resistance when flowing in hydraulic fracturing fractures and HDR with low permeability. In contrast, for EGS-E, 1# working fluid flows in the tunnel and the large fractures formed by gravel accumulation with minimal flow resistance, thus significantly reducing frictional energy consumption in operation.

On the other hand, the EGS-D requires a significant amount of energy to lift the working fluid from the bottom of the production well to the surface. For EGS-E, using the thermosiphon principle, the 2# working fluid can be self-circulating within the closed pipe, minimizing the energy consumption associated with lifting the fluid.

Good safety and environmental friendliness. The high level of safety of EGS-E is reflected in the high level of its own safety and the low risk to the outside environment. In terms of its own safety, considering the convenience of tunnel construction, tunnel support methods such as wet shotcrete support, bolt support, etc. can be used to reinforce tunnels and shafts in the EGS-E structure. As EGS-E is structured in a distributed manner, just as you cannot put all your eggs in one basket, even a local disruption will not affect the whole system too much. During the repair of partial damage, other intact systems can continue to operate without stopping work.

In terms of low risk to the outside environment, EGS-E does not induce earthquakes as it does not require hydraulic fracturing. The mining industry has triggered small micro-earthquake events during its development, but large-scale earthquakes have never been reported. The tunnel walls are sealed by a low-permeability coating so that working fluids do not leak and therefore do not pollute the groundwater.

Controllable financial risk. The EGS-E has the advantages mentioned above, so that the amount of heat that can be extracted and the amount of revenue that can be obtained can be quantified with a high degree of accuracy, which is not possible for the current EGS-D, which is highly stochastic. The geothermal reservoir is innovatively divided into the near-field heat extraction zone and the far-field heat extraction zone, where the revenue generated from the near-field heat extraction zone can at least pay back the capital and the far-field heat extraction zone can be used to create net profits; therefore, the method can bring a definite return to the investor and has good prospects for commercial application.

4.3.5 Challenges

In the above introduction, we have sketched out the blueprint for EGS-E, but how to implement it is the question we need to face, such as excavating tunnels in high temperature and ultra-deep rock masses and maintaining their long-term stability; the efficient transfer of heat from underground to the surface.

The Mponig gold mine reaches a depth of over 4,350 m, which shows us that it is possible to build tunnel structures kilometres underground. However, the rock temperature at this mine is only 60℃, whereas HDR often exceeds 200℃. Consequently, deep excavation remains a challenge, especially in the presence of ultra-high temperatures. Recently, the German company Herrenknecht has developed a shaft boring machine that can drill tunnels with a diameter of 7–12 m in hard rock formations with an in situ stress level of up to 120 MPa. This gives us inspiration. Intelligent mining robots are rapidly iterating and evolving and can be expected to gradually replace humans for deep, high-temperature underground work as their accuracy and power increase [20].

In 2019, Eavor's project at Rocky Mountain House in Alberta, Canada, demonstrated the effectiveness of heat transfer based on the thermosiphon principle [21] (Figure 4.10). The depth of the borehole reached 2,400 m, and the working fluid was water. But EGS-E projects are often built up to 4,000 m below the ground, and finding a better working fluid will be of paramount importance in order to achieve heat transfer at ultra-high differences [22].

It is clear from the above challenges that the challenges faced by EGS-E are not essentially associated with EGS or rock engineering but have more to do with materials, mechanics, artificial intelligence, etc. Therefore, the future development of EGS-E requires the integration of multidisciplinary knowledge systems and the development of a large number of scientific research projects and technological innovations [23].

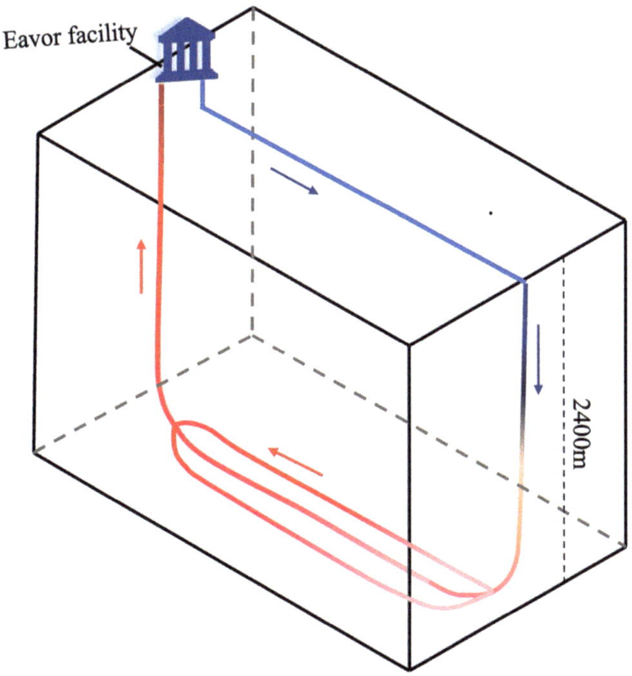

Figure 4.10 Diagram of Eavor's project.

4.4 CONCLUSION

In this chapter, we proposed a new concept named EGS-E as an improvement to the current HDR geothermal extraction method of EGS. For the EGS-E, it uses mining technology to form tunnel systems with specific structures in intact HDR, thus completing the construction of geothermal reservoirs that can be further exploited. The tunnel was used as a space for heat extraction and exchange. The EGS-E uses a two-stage heat exchange system of "fluid-rock (tunnel heat exchange subsystem) and fluid-fluid (pipe heat exchange subsystem)." The EGS-E has many advantages, such as large thermal storage volumes and high outlet temperatures, but the biggest advantage is that it is very controllable. Almost all risks can be controlled by optimizing the design of the structure and adjusting the operating parameters. The core operating principle of EGS-E has been proven by Eavor's project in Alberta, but the construction of geothermal reservoirs is still problematic with current technology. Therefore, the future requires not only the optimization of the structure and operation processes of EGS-E but, more importantly, a multidisciplinary effort.

REFERENCES

[1] Kang F.C., Tang C.A. Overview of enhanced geothermal system (EGS) based one excavation in China. *Earth Science Frontiers*, 2020, 27 (1): 185–193.

[2] Xu T.F., Zhang W. Enhanced geothermal systems: International developments and China's prospects. *Petroleum Science Bulletin*, 2016, 01: 38–44.

[3] Pang Z.H., Luo J., Cheng Y.Z., et al. Evaluation of geological conditions for the development of deep geothermal energy in China. *Earth Science Frontiers*, 2020, 27 (1): 134–151.

[4] Wang G. S., Song X. Z., Yu S., et al. Heat extraction analysis of a novel multilateral-well coaxial closed-loop geothermal system. *Renewable Energy*, 2021, 163: 974–986.

[5] Zhao J., Tang C.A., Wang S.J. Excavation based enhanced geothermal system (EGS-E): Introduction to a new concept. *Geomechanics and Geophysics for Geo-Energy and Geo-Resources*, 2020, 6: 1–7.

[6] Lu S.M. A global review of enhanced geothermal system (EGS). *Renewable and Sustainable Energy Reviews*, 2018, 81: 2902–2921.

[7] Tang M., Li H., et al. Study on preliminarily estimating performance of elementary deep underground engineering structures in future large-scale heat mining projects. *Geofluids*, 2019, (4): 1–10.

[8] Asai P., Panja P., Mclennan J., et al. Performance evaluation of enhanced geothermal system (EGS): Surrogate models, sensitivity study and ranking key parameters. *Renewable Energy*, 2018, 122 (JUL.): 184–195.

[9] Mao X., Guo D.B., Luo L., et al. The global development process of hot dry rock (enhanced geothermal system) and its geological background. *Geological Review*, 2019, 65 (6): 1462–1472.

[10] Breede K., Dzebisashvili K., Liu X. L., et al. A systematic review of enhanced (or engineered) geothermal systems: Past, present and future. *Geothermal Energy*, 2013, 1 (1): 4.

[11] Sigfússon B., Uihlein A. *Technology, Market and Economic Aspects of Geothermal Energy in Europe*. European Commission-Joint Research Centre (JRC) Geothermal Energy Status Report, Petten, 2015.

[12] Tester J. W., Anderson B. J., Batchelor A. S., et al. *The Future of Geothermal Energy-Impact of Enhanced Geothermal Systems (EGS) on the United States in the 21st Century*. Cambridge: MIT Massachusetts Institute of Technology, 2006.

[13] Lin W.J., Wang G.L., Shao J.L., et al. Distribution and exploration of hot dry rock resources in China: Progress and inspiration. *Acta Geologica Sinica*, 95 (5): 1366–1381.
[14] Gan H. N., Wang G. L., Lin W. J., et al. Research on the status quo of environmental geology impact of enhanced geothermal system and countermeasures. *Journal of Geomechanics*, 2020, 26 (2): 211–220.
[15] Wyborn D. Update of development of the geothermal field in the granite at Innamincka, South Australia. *Proceedings World Geothermal Congress*, Bali, Indonesia, 2010, pp. 25–30.
[16] Jung R. EGS: Goodbye or back to the future. *ISRM International Conference for Effective and Sustainable Hydraulic Fracturing*, Brisbane, Australia, May 2013, 95–121.
[17] Guo P., Wu B., Li P., et al. Research progress on hydrofracture and evaluation of enhanced geothermal system. *Resources Environment & Engineering*, 2020, 34 (2): 256–265.
[18] Kim K.H., Ree J.H., Kim Y., et al. Assessing whether the 2017 Mw 5.4 Pohang earthquake in South Korea was an induced event. *Science*, 2018, 360 (6392): 1007–1009.
[19] Whetten J.T., Dennis B.R., Dreesen D.S., et al. The US hot dry rock project. *Geothermics*, 1987, 16 (4): 331–339.
[20] Li J.Y., Li H., Zhu Z.M., et al. Numerical study on damage zones induced by excavation and ventilation in a high-temperature tunnel at depth. *Energies*, 2021, 14: 1–20.
[21] Yuan W., Chen Z., Grasby S. E., et al. Closed-loop geothermal energy recovery from deep high enthalpy systems. *Renewable Energy*, 2021, 177: 976–991.
[22] Pruess K. On production behavior of enhanced geothermal systems with CO_2 as working fluid. *Energy Conversion and Management*, 2008, 49 (6): 1446–1454.
[23] Cai M., Brown E. T. Challenges in the mining and utilization of deep mineral resources. *Engineering*, 2017, 3 (4): 432–433.

Chapter 5

Geothermal reservoirs

Kumari W.G.P.
University of Wollongong

P.G. Ranjith
Monash University

CONTENTS

5.1 INTRODUCTION

Geothermal energy can be simply identified as the energy stored in the earth. There is a steady flow of heat from the center of the earth through the earth's surface into the atmosphere. The decay of radioactive isotopes, including uranium (U^{238} and U^{235}), thorium (Th^{232}) and potassium (K^{40}), found in different constituent rock types, generates this heat. Radioactive heat production can be calculated using an empirical equation proposed by Rybach (1976), expressed as:

$$A\left(\mu W/m^3\right) = \rho(0.0952C_U + 0.0256C_{Th} + 0.0348C_K \tag{5.1}$$

where A = radioactive heat, ρ = density of rock, and C_U, C_{Th}, and C_K are the concentrations of uranium (ppm), thorium (ppm) and potassium (%), respectively.

It has been estimated that the total flow of heat from the earth by conduction, convection and radiation is 42×10^{12} W. Of this heat, 8×10^{12} W is produced from the crust, 32.3×10^{12} W from the mantle and 1.7×10^{12} W from the core (Stacey and Loper, 1988).

DOI: 10.1201/9781003271475-5

5.1.1 Hydrothermal systems

Although geothermal energy is found everywhere around the world, its current utilization is limited to places where the geothermal gradient (heat flux) is very high and geological formations exist with permeable reservoir rocks. Since the temperature at the base of the crust is about 1,100°C, the average conductive temperature gradient is considered to be about 31°C/km. However, good geothermal prospects are regarded as being when the geothermal gradient is over 50°C/km (Beardsmore et al., 2010). In addition to a large heat source, hot permeable rocks are essential to making a geothermal resource commercially viable since heat is extracted from a circulating fluid. In addition, the system requires an overlying layer of impervious rock to minimize the dissipation of formation pressure and temperature, a reliable recharge mechanism to eliminate depletion of the heat resource, and a supply of water. These five features make a hydrothermal geothermal resource known as a conventional geothermal system, as presented in Figure 5.1. Therefore, to date, the major geothermal plants in the world are located near tectonic boundaries, volcanoes and locations where the crust is fractured.

In hydrothermal systems, the heat of the earth is carried to the surface by convective circulation, and energy comes in either vapor-dominant or liquid-dominant forms. Volcanic systems are the most common type of hydrothermal resource worldwide. Other hydrothermal resources can be identified, such as porous sedimentary aquifers, which produce higher-temperature resources than surrounding formations due to their low thermal conductivity. This includes sandstones or conglomerates, fractured and/or cavernous rocks such as limestones, and fault systems, such as, sedimentary aquifers such as those found in the Upper Rhine Valley in Europe (Goldstein, 1986;

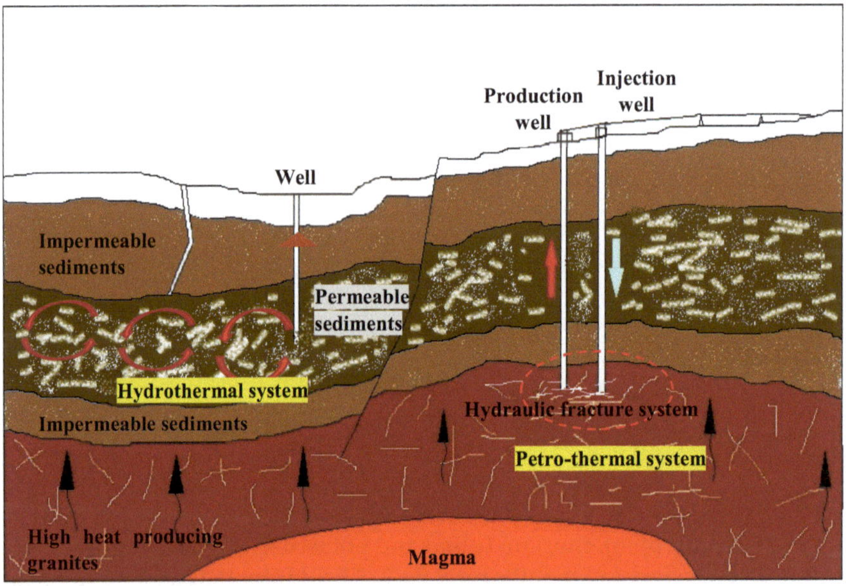

Figure 5.1 Schematic representation of different geothermal systems. Modified and redrawn after Geoscience Australia, 2018.

Huenges and Ledru, 2011). In addition, geo-pressurized systems are also considered hydrothermal systems where warm fluids are significantly confined by surrounding rock and radiogenic resources and groundwater is heated due to the decay of radioactive isotopes. Localized heating increases the average geothermal gradient in all these situations, providing hot water at economical drilling depths.

The development of a hydrothermal system includes the drilling of wells for exploration and production purposes. Generally, reservoir stimulation (mainly hydraulic fracturing) is not conducted in these systems since the formations are highly permeable and geothermal fluids are readily available. However, in some situations, chemical stimulation is performed to enhance near-wellbore flow performance in these systems (Portier et al., 2007). During well drilling and subsequent stimulation, it is crucial to minimize the risk of induced seismicity induced by hydro-mechanical shearing (Kang et al., 2019; Rathnaweera et al., 2020). The thermal energy extracted from these systems is utilized for electricity generation through steam turbine generators. Apart from electricity generation, direct use of geothermal energy, such as for space heating, geothermal heat pumps, or recreational activities, is a common application associated with conventional geothermal systems.

5.2 ENHANCED GEOTHERMAL SYSTEMS

The number of economically attractive geothermal locations is restricted due to the low permeability of reservoir rock and, therefore, the low production rates. These systems are known as petro-thermal systems because only a natural heat source exists, an underground heat exchanger must be artificially created, and water must be supplied for water circulation due to the low formation porosity and the presence of small amounts of fluid. Generally, petro-thermal systems are identified as having hot (>150°C) and dry crystalline or dense sedimentary rocks with natural permeability less than $10^{-14}m^2$. Hence, before stimulation, the productivity index (the ratio of the flow rate to the pressure drawdown; a measure indicating the ability of a well to produce) is less than $10^{-2}m^3/(MPa)$, which means that economic production cannot be achieved without stimulation (Breede et al., 2013; Kumari and Ranjith, 2019).

Engineering techniques are adapted for geothermal resources where adequate fluid circulation and fracture permeability are not present in petro-thermal systems. These include enhancing the permeability of the rock by different stimulation technologies (hydraulic, thermal, and chemical stimulation) to enlarge the extent of the productive geothermal field, improving drilling and reservoir assessment technology, improving the thermodynamic cycle, introducing supercritical fluid systems, and improving exploration methods for deep geothermal reservoirs (Ranjith et al., 2020; Tester et al., 2006). Therefore, these engineered geothermal systems are considered unconventional geothermal resources and are known as enhanced or engineered geothermal systems (EGSs).

5.2.1 Different types of EGSs

Different types of EGSs can be identified depending on the hydro-geological conditions of geothermal reservoirs. In 1970, the initial hot dry rock (HDR) concept was introduced in a patent describing a geothermal system consisting of hot and dry rock as a heat source, and the heat exchanger had to be created artificially (Potter et al., 1974). An HDR reservoir is accessed by drilling a well into a body of crystalline rock and pumping in high-pressure water to open the natural joints of the rock and create artificial fractures. As discussed earlier, these HDRs are typically crystalline rocks heated because of upward convection and conduction of heat from the earth's mantle and core or the decay of radioactive elements in the crust. At the reservoir depths considered in this concept, HDRs primarily consist of granite with a high temperature but very low intact permeability and a small amount of stored fluid, as implied by the terminology "hot dry rocks." However, during the history of deep drilling, it was observed that most rocks are actually not completely dry but contain some naturally occurring water stored in existing fractures or faults. This led to the concept of hot wet rock (HWR), and several geothermal fields in Italy, France, Japan, Iceland, and the USA have already implemented this technology (Baria et al., 1999; Zhang and Zhao, 2020).

In addition, the term engineered geothermal system extends to hot sedimentary aquifers (HSAs), which primarily consist of sedimentary rocks. Elevated geothermal gradients can be found in these systems due to residual heat from old volcanic centers and radioactive decay in basement rocks, which act as heat sources for the overlying aquifers. In addition, some HSA reservoirs formed due to rapid tectonic uplift, which brought a deep hot water formation closer to the surface (Breede et al., 2015). The newly developing supercritical volcanic geothermal systems, which utilize a magma body through deep drilling in the brittle-ductile zone are also considered EGSs. These ultra-high-temperature geothermal systems are mainly associated with active volcanic centers with high geothermal gradients and high heat flows. However, due to the technical difficulties of deep drilling at extreme temperatures, only a few wells can achieve supercritical water conditions, which can be acquired at temperatures of 374°C and pressures of 22.1 MPa (Reinsch et al., 2017). The Beyond Brittle Project in Japan, the Iceland deep drilling project, the Newberry deep drilling project in the USA, the DESCRAMBEL project in Italy, the Hotter and Deeper project in New Zealand, and the GEMex project in Mexico are examples of major ongoing supercritical volcanic geothermal projects (Dobson et al., 2017; Reinsch et al., 2017). Table 5.1 summarizes typical parameter ranges for the different EGSs discussed.

5.2.2 State of the art

The reservoir development process of a typical EGS includes reservoir stimulation to enhance the permeability of the rock, as depicted in Figure 5.1. A hydraulically

Table 5.1 Typical parameter ranges for different EGSs (Breede et al., 2013; Dobson et al., 2017)

Parameter	*Petro-thermal*	*Hot sedimentary aquifer*	*Supercritical volcanic*
Permeability	10^{-19}–10^{-14} m^2	10^{-15} to 10^{-12} m^2	Unknown
Porosity	0.01–0.25	0.08–0.28	Unknown
Temperature	130°C–400°C	76°C–171°C	>374°C
Well depth	1,800–5,000 m	1,280–4,450 m	3,500–5,000 m
Rock type	Igneous	Sedimentary	Igneous
Stimulation method	Hydraulic	Hydraulic/chemical	Hydraulic/thermal

pressurized liquid is pumped with proppants such as sand, ceramic, chemicals and gel to prevent the closure of cracks when water pressure is removed after injection of the fluid. This process is referred to as hydraulic fracturing (Li et al., 2022). The injected water is then heated by heat exchange in the HDRs, and subsequently, one or more production wells are drilled in the reservoir zone to transport heat to the surface. With thermal energy, liquids with a low boiling point are heated and turned into steam to drive steam turbines for power generation. During this operation, fluid circulates around a closed loop, and the system has the potential for steady, long-term energy production.

High flow rates and the thermodynamic efficiency of the circulation fluid determine the economics of the EGS in this process. Therefore, the creation of an interconnected fracture network in the HDR is crucial, allowing heat transportation by circulating the heat-transfer fluid and maintaining sufficient residence time (Tester et al., 1989; White et al., 2018). Appropriate reservoir stimulation techniques need to be employed to enhance the permeability of the reservoir for effective circulation of the fluid between injection and production wells. Different reservoir stimulation methods are employed in the industry, including hydraulic, chemical, and thermal stimulation. Table 5.2 summarizes selected major EGS development projects, including the reservoir characteristics and the major stimulation techniques employed. The following subsections provide a brief overview of different stimulation techniques, including the fundamental mechanisms, state-of-the-art technology with selected case studies, and research directions for future developments.

5.3 HYDRAULIC STIMULATION

Hydraulic stimulation, or hydraulic fracturing, is commonly used to enhance permeability in tight-gas formations and has been adopted to stimulate geothermal reservoirs, particularly EGSs. Permeability is a critical parameter in geothermal power extraction because, considering the steady-state Darcy equation of fluid flow, permeability is proportional to the rate of production of a geothermal reservoir (Zarrouk and Moon, 2014):

$$q = \frac{2\pi kh}{s\mu B}\left(p_r - p_w\right) \tag{5.2}$$

where q is the production rate, k is the permeability, h is the thickness of corresponding strata, μ is the fluid viscosity, s is the skin, p_r is the reservoir pressure, p_w is the well-flowing pressure, and B is the formation volume fracture (the ratio of reservoir volume to production volume).

5.3.1 Hydraulic fracturing process

Hydraulic simulation involves the injection of pressurized fluid that exceeds the fracturing pressure of the reservoir. The increased pressure in the naturally fractured rock mass reduces the effective normal stress acting across the natural fractures, which promotes shearing. Shearing-induced dilation further results in the enhancement of the permeability of the reservoir (Gischig and Preisig, 2015). However, this self-propping

Table 5.2 Selected major EGS developments

Project	*Rock formation*	*Permeability* (m^2) *or transmissivity* $(m^2 s^{-1})$	*Max well depth (km)*	*Max temperature reached (°C)*	*Stimulation techniques employed*	*Output flow rate (kg/s)*	*Output temperature (°C)*	*References*
Fenton Hill, USA	Granite		2.6	185	Hydraulic fracturing	7–16	135–140	Brown (1995), Tester et al. (2006)
			4.39	327	Hydraulic fracturing; acidizing technique with Na_2CO_3 to remove quartz.	5–6	183–190	
Northwest Geysers, USA			3.4	400	Hydraulic fracturing; thermal stimulation is used by injecting cold water into hot rock.			Rutqvist et al. (2015)
Rosemanowes, UK	Granite	$K = 10^{-18}$ to 10^{-17}	2.65	<100	Hydraulic fracturing with proppants including viscous gel stimulation	4–15		Parker (1999), Tester et al. (2006)
Fjällbacka, Sweden	Granite	$K = 10^{-18}$ to 10^{-17}, $T = 10^{-8}$ to 10^{-7}			Hydraulic fracturing with water and viscous gel including proppants; acidizing with HCl and HF and proppant in the second phase			Wallroth et al. (1999)
Bad Urach, Germany	Gneiss	$T = 10^{-7}$ to 10^{-6} rock matrix, 10^{-4} to 10^{-3} fractures	4.45	170	Hydraulic fracturing	50	170	Breede et al. (2013), Sigfússon and Uihlein (2015)

(Continued)

Table 5.2 (Continued) Selected major EGS developments

Project	*Rock formation*	*Permeability (m^2) or transmissivity ($m^2 s^{-1}$)*	*Max well depth (km)*	*Max temperature reached (°C)*	*Stimulation techniques employed*	*Output flow rate (kg/s)*	*Output temperature (°C)*	*References*
Unterhaching, Molasse, Germany			3.59	134	Hydraulic fracturing	150		DiPippo (2012), Wolfgramm et al. (2007)
Ogachi, Japan	Granodiorite	$K = 8 \times 10^{-16}$ to 2×10^{-14}	1	228	Hydraulic fracturing; experiments on CO_2 sequestration with CO_2 injection	0.67	127.5	Kaieda et al. (2005)
Hijiori, Japan	Granodiorite	$K = 10^{-19}$ to 10^{-21}	2.3	270	Hydraulic fracturing	15	150–200	Matsunaga et al. (2005)
Soultz-sous-Forêts, France	Granite	$K = 4 \times 10^{-19}$	5	200	Hydraulic fracturing and acidizing with HCl and HF	31–35	175	Genter et al. (2010)
Cooper Basin, Australia	Granite		4.91	279	Hydraulic fracturing	15–27	213	Geodynamics (2014), Mills and Humphreys (2013)
Paralana, Australia	Meta-sediments, granite		4.01	176	Hydraulic fracturing			Bahadori et al. (2013)
Basel, Switzerland	Granite	$K = 10^{-17}$	4.63	175	Hydraulic fracturing and massive water injection resulted in a high frequency of micro-seismic events with a magnitude of 2.6 and 3.4.		150–200	Breede et al. (2013), Häring and Hopkirk (2002)

mechanism is not always sufficiently efficient (Shiozawa and McClure, 2014). Therefore, to maintain a higher injection rate than the rate of fluid loss through the created fractures, appropriate proppant types are generally employed to prop open the created fractures after releasing the applied pressure (Liang et al., 2016; Zhang et al., 2022).

The typical hydraulic fracturing process can be divided into several steps (Cuss et al., 2015): (i) perforation or cementation and lining of the well to isolate the host rock; (ii) isolation of the perforated section using a packer; (iii) injection of high-pressure fluid, proppants and other additives into the packed section such that the pore pressure exceeds the tensile strength of the formation, causing the initiation of the hydraulic fracture; (iv) flushing or continuing the injection to remove excess proppants; and (v) removal of packers to enable flow-back of the fracturing fluid. A wide range of fracturing fluids are available, including water, foam, oil, acid, alcohol, and other substances; however, water-based fluids are widespread in EGS applications due to the challenging high-temperature conditions of the geothermal reservoir. Different additives, including proppants, acids, biocides, crosslinkers, corrosion inhibitors, scale inhibitors, friction reducers, gelling agents, and surfactants, can be combined to enhance the performance of hydraulic fracturing (Barati and Liang, 2014; Sutra et al., 2017). However, the performance of additives in geothermal applications must be examined at high temperatures.

5.3.2 Breakdown curve

A typical pressure-time curve or breakdown curve of a hydraulic fracturing operation is illustrated in Figure 5.2. As shown in the figure, after the well-bore pressure reaches fracture initiation, followed by the breakdown pressure, the fracture pressure drops to the fracture propagation pressure. Once the maximum length of the fracture under a

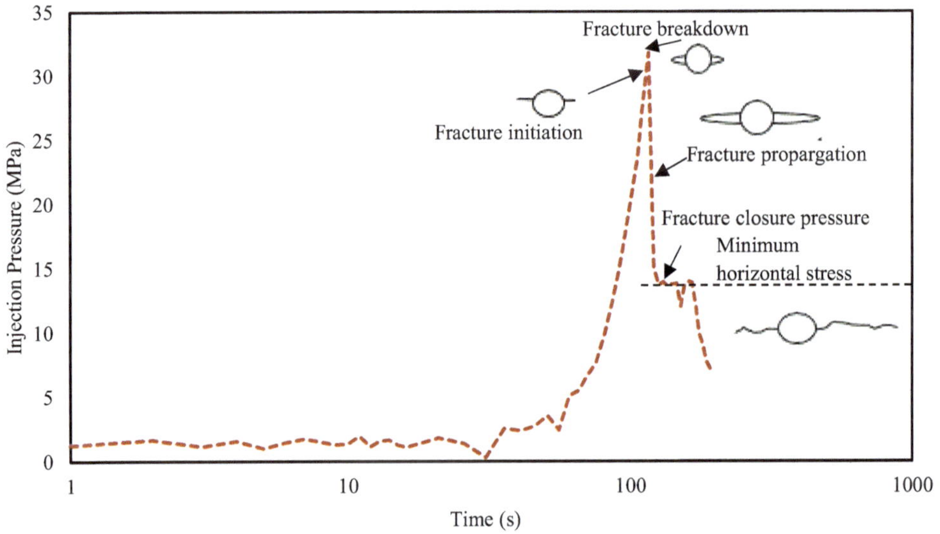

Figure 5.2 A typical pressure-time curve during a hydraulic fracturing experiment without proppant injection. Data: Kumari et al., 2018a.

particular condition is reached (i.e., fluid leak-off in permeable formations), fracture propagation stops (Sarmadivaleh and Rasouli, 2015). Estimation of breakdown pressure and fracture propagation pressure and understanding the fracture geometry are necessary for any hydraulic fracturing operation. However, these depend on a number of parameters, including reservoir properties (in-situ stress state, temperature, mechanical characteristics, and mineralogy), fracturing process parameters (injection flow rate fluid injection method – continuous or intermittent, and stage of fracturing), the properties of the fracturing fluid (density, viscosity, and compressibility), the quantity and quality of proppants and other required additives, and the characteristics of discontinuities (presence of natural fractures, bedding planes, and associated heterogeneity) (Kumari et al., 2018a).

Several theories have been developed to estimate the fracture breakdown pressure. The well-known Hubbert and Wills (HW) criterion (Eq. 5.3) is based on elastic theory for impermeable rocks (Hubbert and Willis, 1957), while the Haimson and Fairhurst criterion (Eq. 5.4) is based on poroelastic theory (Haimson and Fairhurst, 1967). The equation considers the poroelastic stress induced by fluid permeation into rocks. This equation has been extended to include the characteristics of the bore during pressurization by Detournay and Cheng, as shown in Eq. 5.5 (Detournay, 2004; Detournay and Cheng, 1992):

$$P_b = 3\sigma_h - \sigma_H + \sigma_T - P_0 \tag{5.3}$$

$$P_b = \frac{3\sigma_h - \sigma_H + \sigma_T - P_0}{2(1-\eta)} \tag{5.4}$$

$$P_b = \frac{3\sigma_h - \sigma_H + \sigma_T - P_0}{1+(1-2\eta)h(\gamma)} \tag{5.5}$$

where P_b is the breakdown pressure, σ_h is the minimum horizontal stress, σ_H is the maximum horizontal stress, σ_T is the tensile strength of the formation, P_0 is the initial pore pressure, and η and γ are the poroelastic parameter and dimension pressurization rate, respectively, which is defined as follows:

$$\eta = \frac{\alpha(1-2v)}{2(1-v)} \quad 0 \le \eta \le 0.5 \tag{5.6}$$

$$\gamma = \frac{A\lambda^2}{4cS} \quad 0 \le \lambda \le \infty \tag{5.7}$$

where α is the Biot parameter, v is Poisson's ratio, A is the borehole pressurization rate, λ is the micro-crack length scale, c is the diffusivity coefficient, and S is the stress.

It is essential to consider thermal stresses when evaluating the breakdown pressure in a geothermal environment. According to Stephens and Voight (1982), the breakdown pressure of impermeable formations under thermal stresses due to the injection of cold water can be estimated from:

$$P_b = 3\sigma_h - \sigma_H + \sigma_T - P_0 - \frac{\alpha E \Delta T}{1-v} \tag{5.8}$$

where T is the temperature and α is the thermal expansion coefficient.

The breakdown pressure under geothermal conditions is generally affected by many factors, including the in-situ stresses, the orientation of the well, the borehole diameter, the properties of the rock, including tensile strength, elastic modulus and permeability, the injection flow rate, the cooling of the near-well area by the drilling fluid and the capacity of the drilling fluid to build a filter cake and seal the fracture (Kumari et al., 2018a; Zhang, 2019). Therefore, the simplifications and assumptions employed to derive theoretical equations should be noted, and appropriate field techniques, for example, diagnostic fracture injection, are suggested for field applications.

5.3.3 Use of different fracturing fluids

To enhance the efficiency of EGS, it is important to create an interconnected fracture network that enables the extraction of heat from a large surface area of the hot rock. It is essential to maintain sufficient retention time for the circulating fluid and prevent "short-circuiting" to harness heat from the hot rock and maintain the fluid flow at economical production rates during long-term fluid circulation. Due to the greater solvent capacity of water, particularly under high-temperature and high-pressure conditions, mineral dissolution and subsequent precipitation create many challenges for the long-term productivity of the reservoir due to the reduction of fracture permeability with time. Therefore, the scientific community has studied the frackability, fluid flow and heat transfer of different working fluids, and the use of CO_2 in the EGS environment has been considered a viable solution (Singh et al., 2020).

CO_2, which behaves as a supercritical fluid under geothermal conditions (temperature > 31°C and pressure > 7.3 MPa), has been identified as an alternative to water-based fluids due to its favorable thermodynamic properties, including low viscosity, high compressibility, buoyant drive, the low solvent capacity of rock minerals and its environmental merits, including preserving precious water and potential CO_2 sequestration. Figure 5.3 presents comparisons of typical breakdown curves of water- and CO_2-based fracturing for experiments conducted under different temperature and pressure conditions (Isaka et al., 2019; Kumari et al., 2018a). In a particular pressure-time curve, four main phases can be identified: (i) initial pressure development; (ii) well-bore pressurization, where linear pressure increment can be observed; (iii) fracturing, where the well-bore pressure reaches fracture initiation followed by fracture breakdown; and (iv) post-failure without proppant injection. Due to the low compressibility and phase transition of CO_2 from gas to the supercritical state, gradual and delayed initial pressure development can be observed in CO_2-based fracturing. However, well-bore pressurization rates are similar for both water and CO_2. Interestingly, the breakdown pressures under CO_2-based fracturing are 10%–20% lower than those for water-based fracturing, highlighting the ability of CO_2 to percolate into micro-cracks and pores due to its low viscosity. Irrespective of the fracturing fluid, breakdown pressure linearly increases with confining pressure or reservoir depth (or confining pressure), and breakdown pressure linearly decreases with increasing temperature due to the reduction of the tensile strength of rocks with increasing temperature and the enhancement of fluid leak-off due to thermally induced micro-cracks due to the injection of cold fluid into the hot reservoir rock (Kumari et al., 2019).

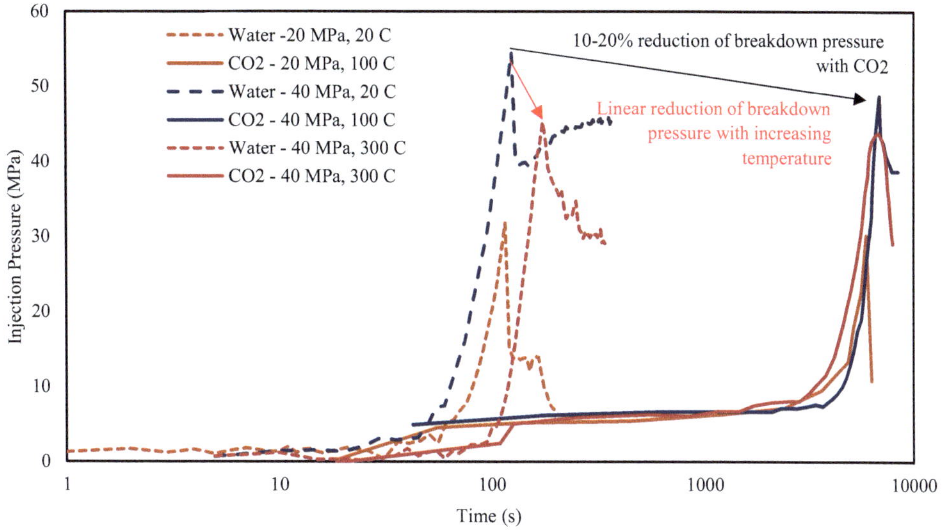

Figure 5.3 Pressure-time plots during water and CO_2 fracturing under different experimental conditions. Data: Isaka et al., 2019; Kumari et al., 2018a.

The morphology of induced hydraulic fractures is affected by the fracturing fluid, mineralogy, and heterogeneity of the reservoir rock and the high-temperature conditions. As presented in Figure 5.4, CT scanning technology can provide valuable insights into hydraulic fracturing mechanisms under geothermal conditions. According to the experimental work of Isaka et al. (2019), due to the greater percolation effect, CO_2-induced fractures create more complex fracture patterns with high tortuosity and fractures with multiple secondary branches. Therefore, the fracture permeability of CO_2-induced fractures is over three orders higher than that of water-induced fractures under similar stress conditions. The injection of cold fluid into hot rock (referred to as thermal stimulation, which is discussed in detail in the next section) can further result in a porous zone close to the wellbore due to thermally induced inter- and intra-crystalline cracks. In addition, since the viscosity of the injection fluid decreases due to temperature, fluids can permeate into micro-cracks, resulting in the induction/extension of hydraulically- or thermally induced micro-cracks. However, experimental evidence indicates that due to enhanced fluid leak-off, fracture apertures are relatively small under high-temperature conditions (Bijay and Ghazanfari, 2021; Kumari et al., 2018a).

5.3.4 Effect of reservoir stress state

The stress state of the formation is the main factor determining fracture orientation and morphology. Reservoir rocks are subject to complicated, heterogeneous stress fields. The vertical stress component is related to the self-weight of the formation, which is partly transmitted to the horizontal stress. Horizontal stresses are created

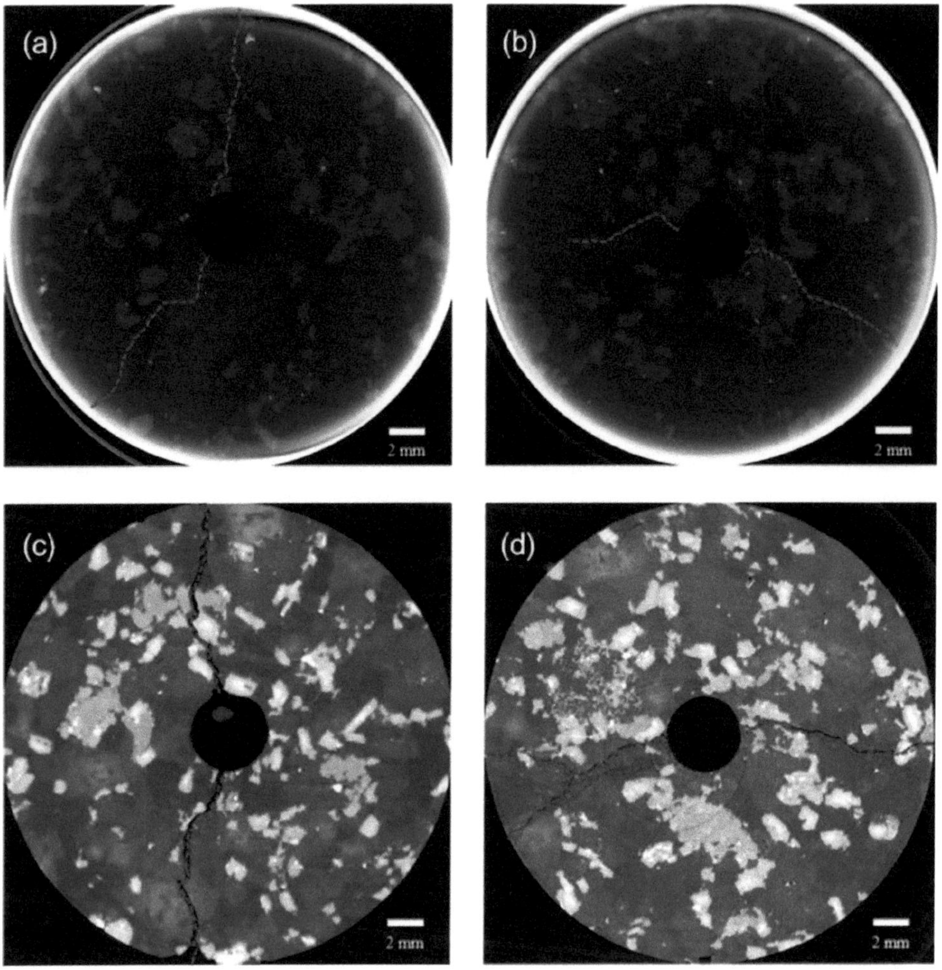

Figure 5.4 Well-bore and propagated hydraulics of granite samples: experiments conducted at (a) 20°C, 40 MPa by water injection; (b) 100°C, 40 MPa by water injection; (c) 50°C, 40 MPa by $ScCO_2$ injection; and (d) 100°C, 40 MPa by $ScCO_2$ injection (Isaka et al., 2019).

by tectonic activity, rock anisotropy, and geological discontinuities (Goodman, 1989). In their experimental work, Zoback et al. (1977) found that a hydraulic fracture is always propagated along the major principal stress. Furthermore, the difference in horizontal stresses determines the breakdown pressure, shape, quantity, and fracture propagation. According to Zhou et al. (2010), a high horizontal stress difference results in random multiple branches, while a low horizontal stress difference results in partly vertical planar fractures with branches. Pre-existing fractures also play a dominant role in the propagation of hydraulic fractures. Hydraulic fractures are capable of crossing, dilating and arresting along natural fractures, depending on the direction of the least-compressive stress, the horizontal stress difference, the angle of intersection and the location of the natural fracture (Blanton, 1982; Guo et al., 2014; Warpinski and Teufel, 1987; Zhou et al., 2010). According to Abass et al. (2009), the near-wellbore stress condition mainly controls the initiation and propagation of hydraulic fractures in the early stages, and once a fracture is extended, its propagation is controlled by the original stress field. Therefore, it is recommended to perforate the well-bore to bypass the near-wellbore stress and create the desired fracture, perpendicular, angular, or longitudinal to the well-bore.

5.3.5 Chemical stimulation

Apart from hydraulic stimulation (water-based fracturing), the injection of acids to dissolve hydrothermal minerals to improve the injectivity of the reservoir, particularly during long-term circulation, is a common reservoir stimulation technique used in geothermal applications. This is often referred to as chemical stimulation. Two general categories of acid treatment exist: matrix acidizing and fracture acidizing. Matrix acidizing is commonly applied in conventional geothermal applications, which employ low injection pressures without introducing induced fractures. This technique is mainly used to reduce formation damage by dissolving rock minerals and etching their surfaces. In contrast, hydraulic fracturing technology and acid injection are combined in fracture acidizing, enabling the creation of new fractures due to fluid pressures above fracture initiation and the dissolution of rock minerals in fractures. Therefore, this technology causes a significant enhancement of fracture permeability by increasing the fracture aperture. The acid types most commonly used include hydrochloric and hydrofluoric acids and their mixtures, as well as organic acids in particular situations. The nature of the injected acids and chemicals differs for different purposes (Portier et al., 2007; Sutra et al., 2017).

5.4 THERMAL STIMULATION

The thermal stimulation technique plays an important role in the enhancement of the near-well productivity of a geothermal reservoir. Injection of cold water into a geothermal reservoir, which causes continuous thermal stresses, results in thermal cracking due to the induced thermal shocks. According to Ghassemi et al. (2007), these thermally induced cracks typically initiate and propagate perpendicular to the hydraulically induced main fracture and can considerably increase the surface area of the fluid flow system and enhance heat transfer. The higher the thermal gradient (or the temperature difference between the reservoir rock and the injectable fluid), the higher the thermal cracking potential. Therefore, this technique has been successfully implemented in high-enthalpy reservoirs in volcanic and metamorphic formations, including at Northwest Geysers, Coso and Newberry (USA) and Larderello (Italy) (Breede et al., 2013). The thermal stimulation technique has been identified as a promising technology for deep geothermal well-drilling projects, including supercritical EGS.

5.4.1 Thermal damage mechanism

The thermal damage mechanism of rocks is reasonably well understood; however, the thermal stimulation mechanism under geothermal environments and the associated production enhancement are not well known as yet due to limited appropriate instrumentation and field data. According to Wong and Brace (1979), with increasing temperature, the internal stress of the rock mass increases as a result of the mismatch between the thermo-elastic coefficients of different minerals. This leads to anisotropic thermal expansion, and once the internal stress exceeds the crack closure pressure, thermal cracks are initiated, as explained in Eq. 5.9.

$$\sigma = E\Delta\alpha\Delta T \tag{5.9}$$

where E, $\Delta\alpha$, and ΔT are the matrix Young's modulus, the difference of the thermal expansion coefficient of distinct minerals, and the temperature difference, respectively.

Other mechanisms associated with thermal stress development include heterogeneous thermal gradients and the thermo-chemical reactions of the rock mass (Dmitriev, 1972; Fredrich and Wong, 1986; Heard and Page, 1982; Wong and Brace, 1979; Yasuhara and Elsworth, 2008). Induced cracks can be propagated along mineral boundaries or the mineral itself, resulting in significant damage to the rock structure. It is generally accepted that inter-crystalline crack density increases in crystalline rock masses at a low-temperature gradient. The temperature gradient increases, the intra-crystalline crack density increases, the fracture aperture increases, and an interconnected fracture network is developed, as depicted in Figure 5.5. However, thermal damage differs depending on the rock type, mineralogy, grain size distribution, and stress condition; as a result, the heterogeneity effect needs to be considered (Shao et al., 2015; Wong, 1982).

5.4.2 Thermo-hydro-mechanical alterations to reservoir rocks

Thermal stimulation can induce significant alterations in the mechanical and physical properties of reservoir rocks, including reduced strength and elastic modulus and increased porosity and permeability. Many studies have investigated the mechanical response of rock at elevated temperatures, and a general negative correlation has been observed for mechanical characteristics of crystalline rocks (compressive strength, tensile strength, and Young's modulus) under high temperatures (Dwivedi et al., 2008; Heuze, 1983; Kumari et al., 2017). Intermittent cold-water injection into hot rock can result in substantial thermal damage, which is critical to ensuring well-bore integrity (Kumari et al., 2017). The tensile strength of the stimulated reservoir is also an important mechanical property since fracture initiation during hydraulic fracturing is generally considered to be a tensile phenomenon. For example, Kumari et al. (2018a) and Isaka et al. (2019) correlated the linear reduction of breakdown pressure with increasing temperature and temperature-dependent tensile characteristics in high-temperature hydraulic fracturing experiments.

Several researchers have observed a decrease in permeability with temperatures ranging from 50°C to 400°C due to the reduction of throat diameter as a result of thermally induced volumetric expansion, followed by a sharp increment of permeability

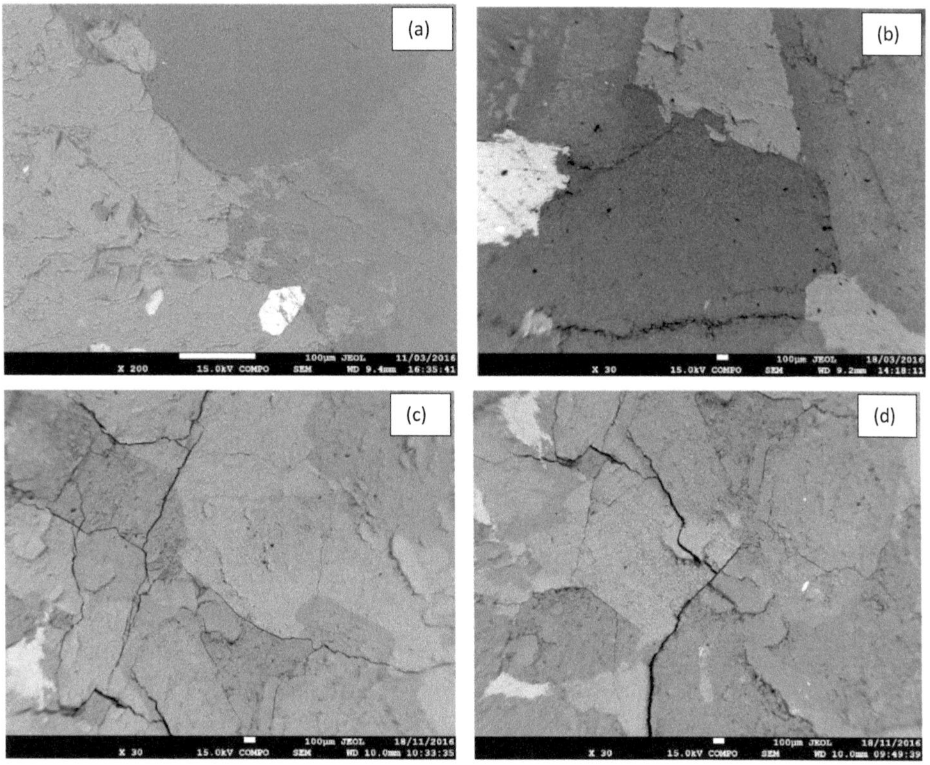

Figure 5.5 Inter- and intra-crystalline thermal cracks of granite samples: samples were heated at (a) 20°C (b) 300°C, (c) 600°C, and (d) 800°C.

at higher temperatures in both intragranular and mixed-crack networks because of increases in both throat diameter and connected porosity volume (Chaki et al., 2008; Kumari et al., 2018b; Wong, 1982). Figure 5.6 depicts an evident correlation between porosity/permeability and thermally induced damage. As a result of the temperature gradient, existing cracks can be extended, new cracks can be formed, and fracture apertures can be increased, emphasizing the continuous increment of flow characteristics with increasing thermal damage. In this regard, the contribution of isolated pores is inadequate for the fluid transport of low-permeable crystalline rocks. Therefore, it is crucial to understand the connected porosity and the throat diameter upon intermittent cold-water injection. Geraud (1994) deduced the permeability of thermally cracked granite with different pore structures and identified that permeability increases linearly with connected pore volume and the square of throat diameter. A later study by Shang et al. (2019) determined the proportions of different pore types based on the pore radius upon thermal damage, as presented in Figure 5.7. Based on the results, the evolution of pore volume in granitic rocks with thermal treatment occurs in several stages. (i) reduction of large-big pore volumes as a result of thermally induced volumetric expansion; (ii) significant increment of volumes of small and micropores due to the development of intergranular and intragranular cracks; (iii) significant increment of large-big pore volumes due to the extension or widening

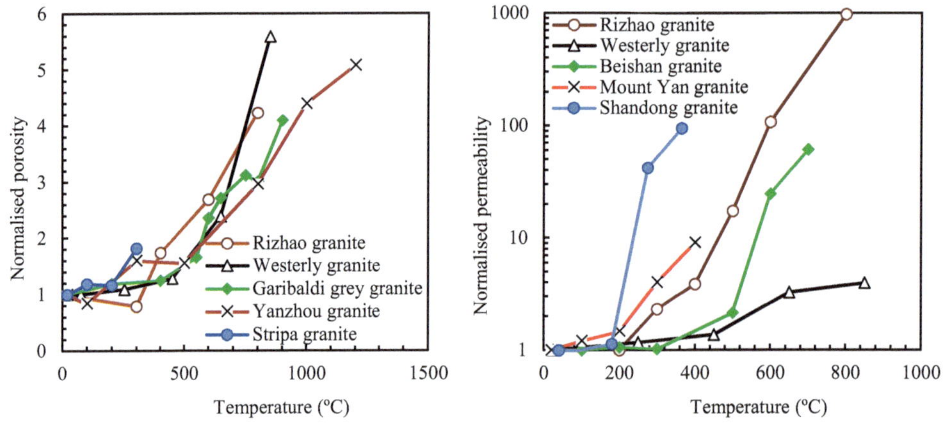

Figure 5.6 Variation of normalized (a) porosity and (b) permeability of different granitic rocks with temperature (Chen et al., 2017; Feng et al., 2018; Griffiths et al., 2017; Nasseri et al., 2009; Shang et al., 2019; Wang et al., 1989; Yu et al., 2014).

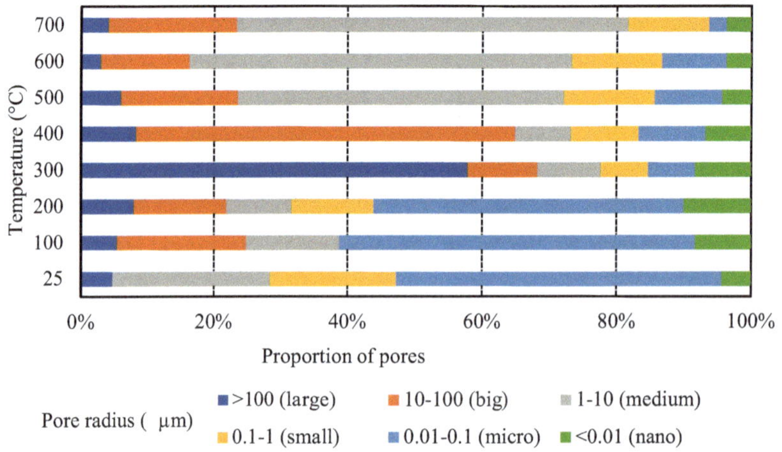

Figure 5.7 Thermal damage mechanism – volume proportion of different pore types in granite under increasing temperature. Data: Shang et al., 2019.

of thermally induced cracks; and (iv) a substantial increment of medium pore volume. This major increment of fracture density is correlated with the volume increase in quartz crystals with the transition of the crystalline structure from α to β, which results in the initiation of new cracks and the propagation of original cracks in the rock matrix (Glover et al., 1995).

While progressive thermal shocks associated with cold-water injection can enhance reservoir permeability, pressurization and depressurization cycles lead to the closure of micro-fractures, hence reducing the permeability of the reservoir. As discussed in the previous section, macro-fractures induced by hydraulic fracturing substantially enhance reservoir permeability, particularly when fractures are slightly offset because of shearing or fine fragments or proppants lodged in the fractures. The contribution of the secondary orthogonal micro-fractures induced by thermal stresses may be insignificant since the micro-fractures are highly sensitive to effective stresses (Eggertsson et al., 2020). Therefore, the coupled effect of hydraulic and thermal stimulations should be considered when simulating a geothermal reservoir. The stress state of the reservoir should be critically considered in order to generate fractures orthogonal to primary local faults and fractures typically at a low angle to principal stresses. In addition, the use of appropriate proppants consistent with the fracture aperture is critical to keeping the fractures open to facilitate fluid flow in the reservoir.

5.5 SYNOPSIS

Despite the enormous potential to generate power using geothermal energy, current commercial production is limited to conventional hydrothermal geothermal systems, typically consisting of naturally fractured or porous geological formations with high geothermal gradients. In contrast, technologies are steadily being developed to harness geothermal energy from unconventional geothermal resources, including HDR systems, HSAs and supercritical volcanic systems. It is essential to adopt engineering techniques to boost the production of these systems since they typically have low matrix permeability and a small amount of stored fluid. Therefore, unconventional geothermal resources are often referred to as enhanced or engineered geothermal systems (EGSs).

Enhancement of the permeability of a reservoir by developing an artificial fluid flow system, often known as reservoir stimulation, is a distinguishing feature of EGSs compared to traditional hydrothermal systems. Commonly adopted reservoir stimulation techniques include hydraulic, chemical, and thermal stimulation.

The hydraulic stimulation technology extensively used for fracturing oil and gas wells has been widely adopted in EGSs. Due to the challenges associated with the inherent high-temperature and high-pressure environment of geothermal systems, it is essential to consider the thermal stresses associated with hydraulic fracturing.

Several recent studies have attempted to substitute water-based fracturing fluids with CO_2, which behaves as a supercritical fluid under geothermal conditions. CO_2 has been identified as a promising alternative due to its favorable thermodynamic properties, buoyant drive, low solvent capacity, and CO_2 sequestration merits; however, field trials are required to establish this technology.

Injection of acids, particularly HCl or HF-based acids, has also been shown to enhance reservoir injectivity by dissolving hydrothermal minerals and is often referred to as chemical stimulation. Depending on the injection pressure employed, two technologies can be identified: matrix acidizing and fracture acidizing.

Injection of cold water into hot rock is often referred to as thermal stimulation, and in the EGS environment, hydraulic stimulation is often coupled with thermal stimulation. A higher-temperature differential (or the temperature difference between the injection fluid and hot rock) induces higher thermal stresses, resulting in the initiation and propagation of thermally induced fractures along the weakest planes, including grain boundaries and weaker mineral constituents orthogonal to the hydraulic fractures. However, these micro-fractures are highly sensitive to thermal and effective stresses; as a result, appropriate techniques, including promoting self-propping mechanisms by shearing or injecting appropriate proppant types to keep fractures open, are critical to enhancing the production of reservoirs.

REFERENCES

Abass, H.H., Soliman, M.Y., Al-Tahini, A.M., Surjaatmadja, J.B., Meadows, D.L. and Sierra, L., 2009. Oriented fracturing: A new technique to hydraulically fracture an openhole horizontal well, *SPE Annual Technical Conference and Exhibition*, New Orleans, Louisiana, OnePetro.

Bahadori, A., Zendehboudi, S. and Zahedi, G., 2013. A review of geothermal energy resources in Australia: current status and prospects. *Renewable and Sustainable Energy Reviews*, 21: 29–34.

Barati, R. and Liang, J.T., 2014. A review of fracturing fluid systems used for hydraulic fracturing of oil and gas wells. *Journal of Applied Polymer Science*, 131(16). https://onlinelibrary.wiley.com/doi/epdf/10.1002/app.40735

Baria, R., Baumgärtner, J., Rummel, F., Pine, R.J. and Sato, Y., 1999. HDR/HWR reservoirs: concepts, understanding and creation. *Geothermics*, 28(4): 533–552.

Beardsmore, G.R., Rybach, L., Blackwell, D. and Baron, C., 2010. A protocol for estimating and mapping global EGS potential. *GRC Transactions*, 34: 301–312.

Bijay, K. and Ghazanfari, E., 2021. Geothermal reservoir stimulation through hydro-shearing: an experimental study under conditions close to enhanced geothermal systems. *Geothermics*, 96: 102200.

Blanton, T.L., 1982. An experimental study of interaction between hydraulically induced and pre-existing fractures, *SPE Unconventional Gas Recovery Symposium*, Pittsburg, PA, 16 May 1982 . Society of Petroleum Engineers.

Breede, K., Dzebisashvili, K. and Falcone, G., 2015. Overcoming challenges in the classification of deep geothermal potential. *Geothermal Energy Science*, 3(1): 19.

Breede, K., Dzebisashvili, K., Liu, X. and Falcone, G., 2013. A systematic review of enhanced (or engineered) geothermal systems: past, present and future. *Geothermal Energy*, 1(1): 4.

Brown, D., 1995. The US hot dry rock program-20 years of experience in reservoir testing, *Proceedings of the World Geothermal Congress*, Florence, Italy, pp. 2607–2611.

Chaki, S., Takarli, M. and Agbodjan, W.P., 2008. Influence of thermal damage on physical properties of a granite rock: porosity, permeability and ultrasonic wave evolutions. *Construction and Building Materials*, 22(7): 1456–1461.

Chen, S., Yang, C. and Wang, G., 2017. Evolution of thermal damage and permeability of Beishan granite. *Applied Thermal Engineering*, 110: 1533–1542.

Cuss, R., Wiseall, A., Hennissen, J., Waters, C.N., Kemp, S., Ougier-Simonin, A., Holyoake, S. and Haslam, R., 2015. Hydraulic fracturing: a review of theory and field experience, British Geological Survey Open Report, OR/15/066. 5pp..

Detournay, E., 2004. Propagation regimes of fluid-driven fractures in impermeable rocks. *International Journal of Geomechanics*, 4(1): 35–45.

Detournay, E. and Cheng, A., 1992. Influence of pressurization rate on the magnitude of the breakdown pressure, *The 33th US Symposium on Rock Mechanics (USRMS)*, Santa Fe, New Mexico, June 1992, American Rock Mechanics Association.

DiPippo, R., 2012. *Geothermal Power Plants: Principles, Applications, Case Studies and Environmental Impact*. Butterworth-Heinemann, Oxford, Waltham, MA.

Dmitriev, A.P., 1972. *Physical Properties of Rocks at High Temperatures*. National Aeronautics and Space Administration; For sale by the National Technical Information Service,Springfield, VA.

Dobson, P., Asanuma, H., Huenges, E., Poletto, F., Reinsch, T. and Sanjuan, B., 2017. Supercritical geothermal systems-a review of past studies and ongoing research activities, *42nd Workshop on Geothermal Reservoir Engineering*. Stanford University, Stanford, CA.

Dwivedi, R.D., Goel, R.K., Prasad, V.V.R. and Sinha, A., 2008. Thermo-mechanical properties of Indian and other granites. *International Journal of Rock Mechanics and Mining Sciences*, 45(3): 303–315.

Eggertsson, G.H., Lavallée, Y., Kendrick, J.E. and Markússon, S.H., 2020. Improving fluid flow in geothermal reservoirs by thermal and mechanical stimulation: the case of Krafla volcano, Iceland. *Journal of Volcanology and Geothermal Research*, 391: 106351.

Feng, Z., Zhao, Y., Zhang, Y. and Wan, Z., 2018. Real-time permeability evolution of thermally cracked granite at triaxial stresses. *Applied Thermal Engineering*, 133: 194–200.

Fredrich, J.T. and Wong, T.F., 1986. Micromechanics of thermally induced cracking in three crustal rocks. *Journal of Geophysical Research: Solid Earth*, 91(B12): 12743–12764.

Genter, A., Evans, K., Cuenot, N., Fritsch, D. and Sanjuan, B., 2010. Contribution of the exploration of deep crystalline fractured reservoir of Soultz to the knowledge of enhanced geothermal systems (EGS). *Comptes Rendus Geoscience*, 342(7–8): 502–516.

Geodynamics, 2014. *Habanero Geothermal Project Field Development Plan*, COM-FN-OT-PLN-01166, Revision No: 1.0. Australia Renewable Energy Agency, Australia. https://arena.gov.au/assets/2016/10/Habanero-Geothermal-Project-Field-Development-Plan.pdf

Geoscience Australia, 2018. *Geothermal Geology*. Geoscience Australia, Australia. Available from: http://ga.gov.au/scientific-topics/energy/resources/geothermal-energy-resources/geothermal-geology.

Geraud, Y., 1994. Variations of connected porosity and inferred permeability in a thermally cracked granite. *Geophysical Research Letters*, 21(11): 979–982.

Ghassemi, A., Tarasovs, S. and Cheng, A.-D., 2007. A 3-D study of the effects of thermomechanical loads on fracture slip in enhanced geothermal reservoirs. *International Journal of Rock Mechanics and Mining Sciences*, 44(8): 1132–1148.

Gischig, V.S. and Preisig, G., 2015. Hydro-fracturing versus hydro-shearing: a critical assessment of two distinct reservoir stimulation mechanisms, *13th ISRM International Congress of Rock Mechanics*, Montreal, May 2015. International Society for Rock Mechanics.

Glover, P., Baud, P., Darot, M., Meredith, P., Boon, S., LeRavalec, M., Zoussi, S. and Reuschlé, T., 1995. α/β phase transition in quartz monitored using acoustic emissions. *Geophysical Journal International*, 120(3): 775–782.

Goldstein, N., 1986. Subregional and detailed exploration for geothermal-hydrothermal resources, A report prepared for the US Department of Energy under Contract DE-AC03-76SF00098 by Lawrence Berkeley Laboratory, University of California.

Goodman, R.E., 1989. *Introduction to Rock Mechanics*, vol. 2. Wiley, New York.

Griffiths, L., Heap, M., Baud, P. and Schmittbuhl, J., 2017. Quantification of microcrack characteristics and implications for stiffness and strength of granite. *International Journal of Rock Mechanics and Mining Sciences*, 100: 138–150.

Guo, T., Zhang, S., Qu, Z., Zhou, T., Xiao, Y. and Gao, J., 2014. Experimental study of hydraulic fracturing for shale by stimulated reservoir volume. *Fuel*, 128: 373–380.

Haimson, B. and Fairhurst, C., 1967. Initiation and extension of hydraulic fractures in rocks. *Society of Petroleum Engineers Journal*, 7(03): 310–318.

Häring, M. and Hopkirk, R., 2002. The Swiss deep heat mining project-the Basel exploration drilling. *GHC Bulletin*, 23(1): 31–33.

Heard, H. and Page, L., 1982. Elastic moduli, thermal expansion, and inferred permeability of two granites to 350C and 55 megapascals. *Journal of Geophysical Research: Solid Earth (1978–2012)*, 87(B11): 9340–9348.

Heuze, F., 1983. High-temperature mechanical, physical and thermal properties of granitic rocks—a review, *International Journal of Rock Mechanics and Mining Sciences & Geomechanics Abstracts*, 20, 3–10.

Hubbert, M.K. and Willis, D.G., 1957. Mechanics of hydraulic fracturing. *Transactions of the AIME*, 210(01): 153–168.
Huenges, E. and Ledru, P., 2011. *Geothermal Energy Systems: Exploration, Development, and Utilization*. Wiley-VCH Verlag GmbH & Co. KGaA .
Isaka, B.A., Ranjith, P., Rathnaweera, T., Wanniarachchi, W., Kumari, W. and Haque, A., 2019. Testing the frackability of granite using supercritical carbon dioxide: insights into geothermal energy systems. *Journal of CO_2 Utilization*, 34: 180–197.
Kaieda, H., Ito, H., Kiho, K., Suzuki, K., Suenaga, H. and Shin, K., 2005. Review of the Ogachi HDR project in Japan, *Proceedings of the World Geothermal Congress*. Antalya, 24–29 April 2005, Citeseer.
Kang, J.-Q., Zhu, J.-B. and Zhao, J., 2019. A review of mechanisms of induced earthquakes: from a view of rock mechanics. *Geomechanics and Geophysics for Geo-Energy and Geo-Resources*, 5(2): 171–196.
Kumari, W., Beaumont, D., Ranjith, P., Perera, M., Isaka, B.A. and Khandelwal, M., 2019. An experimental study on tensile characteristics of granite rocks exposed to different high-temperature treatments. *Geomechanics and Geophysics for Geo-Energy and Geo-Resources*, 5(1): 47–64.
Kumari, W. and Ranjith, P., 2019. Sustainable development of enhanced geothermal systems based on geotechnical research–a review. *Earth-Science Reviews*, 199: 102955.
Kumari, W., Ranjith, P., Perera, M., Li, X., Li, L., Chen, B., Isaka, B.A. and De Silva, V., 2018a. Hydraulic fracturing under high temperature and pressure conditions with micro CT applications: geothermal energy from hot dry rocks. *Fuel*, 230: 138–154.
Kumari, W.G.P., Ranjith, P.G., Perera, M.S.A. and Chen, B.K., 2018b. Experimental investigation of quenching effect on mechanical, microstructural and flow characteristics of reservoir rocks: thermal stimulation method for geothermal energy extraction. *Journal of Petroleum Science and Engineering*, 162: 419–433.
Kumari, W.G.P., Ranjith, P.G., Perera, M.S.A., Chen, B.K. and Abdulagatov, I.M., 2017. Temperature-dependent mechanical behaviour of Australian Strathbogie granite with different cooling treatments. *Engineering Geology*, 229: 31–44.
Li, N., Xie, H., Hu, J. and Li, C., 2022. A critical review of the experimental and theoretical research on cyclic hydraulic fracturing for geothermal reservoir stimulation. *Geomechanics and Geophysics for Geo-Energy and Geo-Resources*, 8(1): 1–19.
Liang, F., Sayed, M., Al-Muntasheri, G.A., Chang, F.F. and Li, L., 2016. A comprehensive review on proppant technologies. *Petroleum*, 2(1): 26–39.
Matsunaga, I., Niitsuma, H. and Oikawa, Y., 2005. Review of the HDR development at Hijiori Site, Japan, *Proceedings of the World Geothermal Congress*, Antalya, 24–29 April 2005, pp. 1–5.
Mills, T. and Humphreys, B., 2013. Habanero pilot project—Australia's first EGS power plant, *Proceedings of 35th New Zealand Geothermal Workshop*, 17–20 November 2013 Rotorua, New Zealand, pp. 37–41.
Nasseri, M., Schubnel, A., Benson, P. and Young, R., 2009. Common evolution of mechanical and transport properties in thermally cracked westerly granite at elevated hydrostatic pressure. *Pure and Applied Geophysics*, 166(5): 927–948.
Parker, R., 1999. The Rosemanowes HDR project 1983–1991. *Geothermics*, 28(4–5): 603–615.
Portier, S., André, L. and Vuataz, F.-D., 2007. Review on chemical stimulation techniques in oil industry and applications to geothermal systems. *Engine, Work Package*, 4: 32.
Potter, R., Robinson, E. and Smith, M., 1974. *Method of extracting heat from dry geothermal reservoirs*. Google Patents.
Ranjith, P., Chandrasekharam, D., Abdulagatov, I., Wu, B. and Wanniarachchige, G.P.K., 2020. Guest editorial for the topical collection: sustainable development and utilization of

geothermal systems. *Geomechanics and Geophysics for Geo-Energy and Geo-Resources,* 6, 70. https://doi.org/10.1007/s40948-020-00188-4.

Rathnaweera, T.D., Wu, W., Ji, Y. and Gamage, R.P., 2020. Understanding injection-induced seismicity in enhanced geothermal systems: from the coupled thermo-hydro-mechanical-chemical process to anthropogenic earthquake prediction. *Earth-Science Reviews*, 205: 103182.

Reinsch, T., Dobson, P., Asanuma, H., Huenges, E., Poletto, F. and Sanjuan, B., 2017. Utilizing supercritical geothermal systems: a review of past ventures and ongoing research activities. *Geothermal Energy*, 5(1): 1–25.

Rutqvist, J., Dobson, P.F., Garcia, J., Hartline, C., Jeanne, P., Oldenburg, C.M., Vasco, D.W. and Walters, M., 2015. The northwest Geysers EGS demonstration project, California: pre-stimulation modeling and interpretation of the stimulation. *Mathematical Geosciences*, 47(1): 3–29.

Rybach, L., 1976. Radioactive heat production in rocks and its relation to other petrophysical parameters. *Pure and Applied Geophysics*, 114(2): 309–317.

Sarmadivaleh, M. and Rasouli, V., 2015. Test design and sample preparation procedure for experimental investigation of hydraulic fracturing interaction modes. *Rock Mechanics and Rock Engineering*, 48(1): 93–105.

Shang, X., Zhang, Z., Xu, X., Liu, T. and Xing, Y., 2019. Mineral composition, pore structure, and mechanical characteristics of pyroxene granite exposed to heat treatments. *Minerals*, 9(9): 553.

Shao, S., Ranjith, P.G., Wasantha, P.L.P. and Chen, B.K., 2015. Experimental and numerical studies on the mechanical behaviour of Australian Strathbogie granite at high temperatures: an application to geothermal energy. *Geothermics*, 54: 96–108.

Shiozawa, S. and McClure, M., 2014. EGS designs with horizontal wells, multiple stages, and proppant, *Proceedings of the 39th Workshop on Geothermal Reservoir Engineering*, Stanford, CA.

Sigfússon, B. and Uihlein, A., 2015. *2014 JRC Geothermal Energy Status Report*. Publications Office of the European Union, Luxembourg.

Singh, M., Tangirala, S.K. and Chaudhuri, A., 2020. Potential of CO_2 based geothermal energy extraction from hot sedimentary and dry rock reservoirs, and enabling carbon geo-sequestration. *Geomechanics and Geophysics for Geo-Energy and Geo-Resources*, 6(1): 1–32.

Stacey, F.D. and Loper, D.E., 1988. Thermal history of the Earth: a corollary concerning non-linear mantle rheology. *Physics of the Earth and Planetary Interiors*, 53(1–2): 167–174.

Stephens, G. and Voight, B., 1982. Hydraulic fracturing theory for conditions of thermal stress. *International Journal of Rock Mechanics and Mining Sciences & Geomechanics Abstracts*, 19(6): 279–284.

Sutra, E., Spada, M. and Burgherr, P., 2017. Chemicals usage in stimulation processes for shale gas and deep geothermal systems: a comprehensive review and comparison. *Renewable and Sustainable Energy Reviews*, 77: 1–11.

Tester, J.W., Anderson, B., Batchelor, A., Blackwell, D., DiPippo, R., Drake, E., Garnish, J., Livesay, B., Moore, M.C. and Nichols, K., 2006. *The Future of Geothermal Energy: Impact of Enhanced Geothermal Systems (EGS) on the United States in the 21st Century*. Massachusetts Institute of Technology, Cambridge, MA.

Tester, J., Brown, D. and Potter, R., 1989. *Hot Dry Rock Geothermal Energy—A New Energy Agenda for the Twenty-First Century*. Los Alamos National Lab., Los Alamos, NM,.

Wallroth, T., Eliasson, T. and Sundquist, U., 1999. Hot dry rock research experiments at Fjällbacka, Sweden. *Geothermics*, 28(4–5): 617–625.

Wang, H., Bonner, B., Carlson, S., Kowallis, B. and Heard, H., 1989. Thermal stress cracking in granite. *Journal of Geophysical Research: Solid Earth*, 94(B2): 1745–1758.

Warpinski, N. and Teufel, L., 1987. Influence of geologic discontinuities on hydraulic fracture propagation (includes associated papers 17011 and 17074). *Journal of Petroleum Technology*, 39(02): 209–220.

White, M., Fu, P., McClure, M., Danko, G., Elsworth, D., Sonnenthal, E., Kelkar, S. and Podgorney, R., 2018. A suite of benchmark and challenge problems for enhanced geothermal systems. *Geomechanics and Geophysics for Geo-Energy and Geo-Resources*, 4(1): 79–117.

Wolfgramm, M., Bartels, J., Hoffmann, F., Kittl, G., Lenz, G., Seibt, P., Schulz, R., Thomas, R. and Unger, H., 2007. Unterhaching geothermal well doublet: structural and hydrodynamic reservoir characteristic; Unterhaching, Bavaria (Germany), *European Geothermal Congress*. Citeseer.

Wong, T.-F., 1982. Effects of temperature and pressure on failure and post-failure behavior of Westerly granite. *Mechanics of Materials*, 1(1): 3–17.

Wong, T.-F. and Brace, W., 1979. Thermal expansion of rocks: some measurements at high pressure. *Tectonophysics*, 57(2): 95–117.

Yasuhara, H. and Elsworth, D., 2008. Compaction of a rock fracture moderated by competing roles of stress corrosion and pressure solution. *Pure and Applied Geophysics*, 165(7): 1289–1306.

Yu, W., Bao-lin, L., Hai-yan, Z., Chuan-liang, Y., Zhi-jun, L. and Zhi-qiao, W., 2014. Thermophysical and mechanical properties of granite and its effects on borehole stability in high temperature and three-dimensional stress. *The Scientific World Journal*, 2014: 650683. https://doi.org/10.1155/2014/650683

Zarrouk, S.J. and Moon, H., 2014. Efficiency of geothermal power plants: a worldwide review. *Geothermics*, 51: 142–153.

Zhang, J.J., 2019. Chapter 11 - Geomechanics applications in hydraulic fracturing. In: J.J. Zhang (Ed.), *Applied Petroleum Geomechanics*. Gulf Professional Publishing, Elsevier, Cambridge, MA; Oxford, UK, pp. 441–481. Available from: https://books.google.com.au/books?hl=en&lr=&id=g5adDwAAQBAJ&oi=fnd&pg=PP1&dq=Geomechanics+applications+in+hydraulic+fracturing.+In:+J.J.+Zhang+(Ed.),+Applied+Petroleum+Geomechanics.+Gulf+Professional+Publishing&ots=20lFsSNgJ9&sig=BS9DaYhHVJe_PZ4wn0ktKpaIs9k#v=onepage&q=Geomechanics%20applications%20in%20hydraulic%20fracturing.%20In%3A%20J.J.%20Zhang%20(Ed.)%2C%20Applied%20Petroleum%20Geomechanics.%20Gulf%20Professional%20Publishing&f=false

Zhang, Y., Wu, Y., Teng, Y., Li, P. and Peng, S., 2022. Experiment study on the evolution of permeability and heat recovery efficiency in fractured granite with proppants. *Geomechanics and Geophysics for Geo-Energy and Geo-Resources*, 8(1): 1–19.

Zhang, Y. and Zhao, G.-F., 2020. A global review of deep geothermal energy exploration: from a view of rock mechanics and engineering. *Geomechanics and Geophysics for Geo-Energy and Geo-Resources*, 6(1): 1–26.

Zhou, J., Jin, Y. and Chen, M., 2010. Experimental investigation of hydraulic fracturing in random naturally fractured blocks. *International Journal of Rock Mechanics and Mining Sciences*, 47(7): 1193–1199.

Zoback, M., Rummel, F., Jung, R. and Raleigh, C., 1977. Laboratory hydraulic fracturing experiments in intact and pre-fractured rock. *International Journal of Rock Mechanics and Mining Sciences & Geomechanics Abstracts*, 14(2): 49–58.

Chapter 6

Induced seismicity in EGS

Observations and triggering mechanisms

Silvia De Simone
University of Rennes

Iman Rahimzadeh Kivi
Spanish National Research Council

Ki-Bok Min
Seoul National University

Jonny Rutqvist
Lawrence Berkeley National Laboratory

CONTENTS

DOI: 10.1201/9781003271475-6

6.1 INTRODUCTION

Human-induced seismicity has been observed in several activities, such as reservoir impoundments [1,2], mining [3,4], fluid extraction in the oil and gas industry [5–7], fluid injection [8] for subsurface wastewater disposal [9], CO2 or gas storage and geothermal energy production [10]. In this chapter, we focus on the occurrence of induced seismicity in geothermal applications, and in particular in enhanced geothermal systems (EGSs). Although this seismicity has generally been of low intensity (e.g., magnitude[1] $M < 2$), it has often reached a magnitude large enough to be felt by local inhabitants and cause damage. Avoiding such levels of seismicity is crucial not only to ensure the safety of the infrastructure and the surrounding buildings but also to preserve public confidence in the technology.

The achievement of this objective passes through the observation of the occurrence of induced seismicity, the identification of the leading causes and mechanisms, and the understanding of their spatio-temporal evolution. In the following, we first summarize some of the most relevant cases of induced seismicity in geothermal systems. Successively, we describe the key factors and the physical processes that control the occurrence of induced earthquakes. Afterwards, we review the state of the art in the modeling of induced seismicity and its forecasting.

A preliminary remark is necessary. Some authors (e.g., [11,12]) suggest a distinction between "induced" seismicity and "triggered" seismicity based on the amount of human-induced stress variation that leads to the seismic event. Induced seismic events are driven by large stress variations, i.e., comparable to the in-situ shear stress causing the fault to slip, whereas triggered events are caused by small stress changes compared to the in-situ stress and anticipate the occurrence of a natural earthquake [13,14]. It is, however, difficult to estimate the stress changes at the hypocenters, so this differentiation is often impossible (see also the discussion in Ref. [15]). Without entering into the debate on the effectiveness of this distinction, in this chapter we do not differentiate between triggered and induced seismicity, and we adopt the expression induced seismicity to express seismicity associated with human activities [16]. Similarly, we use the term triggering mechanisms to define the processes that perturb the stress field and potentially drive seismicity.

6.2 OBSERVED SEISMICITY IN EGS

The number of EGS exploitations has increased during the past 40 years, accompanied by the reported cases of induced earthquakes. In most cases, the magnitude of seismic events does not exceed magnitude 2, although larger events have been occasionally recorded (Table 6.1). In the following, we provide a brief review of the cases of induced seismicity associated with geothermal sites. They are classified into two chronological blocks, according to the year in which the event with the maximum magnitude has been recoreded. More comprehensive surveys can be found in Refs. [10,13,17–19].

[1] All magnitudes in this chapter refer to local magnitudes (M_L, Richter scale – see Section 6.3.1) unless otherwise indicated.

Table 6.1 Maximum earthquake magnitudes (M_L) recorded at selected geothermal sites where the seismicity exceeded M2.

Project	*Mmax*	*Year*	*Activity*	*Depth (km)*	*Rock type*
Monte Amiata (Italy)	3.5	1969	Circulation	3.5	Basement
Torre Alfina (Italy)	3.0	1977	Injection	1.7	Sedimentary
Lardarello-Travale (Italy)	3.2	1977	Circulation	2	Sedimentary
Latera (Italy)	2.9	1984	Injection	1.5	Sedimentary
Rosemanowes (UK)	2.0	1987	Circulation	2.6	Basement
Soultz-sous-Forêts (France)	2.6*	2000	Stimul. GPK2	5.1	Basement
Hijiori (Japan)	2.4*	2002	Circulation	2.2	Basement
Bad Urach (Germany)	1.8	2002	Stimulation	4.3	Basement
Soultz-sous-Forêts (France)	2.9*	2003	Stimul. GPK3	5.1	Basement
Cooper Basin (Australia)	3.7*	2003	Stimulation	4.4	Basement
Berlín Field (El Salvador)	4.4*	2003	Stimulation	1.8	Basement
Coso (USA)	2.8	2004	Stimulation	3	Basement
Soultz-sous-Forêts (France)	2.3*	2004	Stimul. GPK4	5	Basement
Salton Sea (CA, USA)	5.1	2005	Circulation	2.5	Sedimentary
Basel (Switzerland)	3.4*	2006	Stimulation	4.8	Basement
Unterhaching (Germany)	2.4	2007	Circulation	3.4	Sedimentary
Landau (Germany)	2.7*	2009	Circulation	3.2	Mixed
Insheim (Germany)	2.4	2010	Stimulation	3.6	Mixed
Paralana (Australia)	2.9*	2011	Stimulation	3.6	Sedimentary
Hellisheidi (Iceland)	4	2011	Drill-stimul.	2.5	Basement
St. Gallen (Switzerland)	3.5	2013	Circulation	4	Sedimentary
The Geysers (CA, USA)	5.0	2016	Circulation	3	Basement
Pohang (South Korea)	5.4*	2017	Stimulation	4.3	Basement

Maximum values for Coso, The Geysers and Salton Sea are seismic event magnitude M_W. Asterisks denote events occurring after the stop of the injection. Basement refers to both crystalline and metamorphic rocks. Mixed refers to cases where the depth of occurred seismicity is not well constrained. References are provided in the text.

6.2.1 From 1970s to 2000

Seismicity associated with geothermal applications has been detected since the first geothermal exploitations in natural hydrothermal anomalies at the Tuscany–Latium geothermal region (Italy) and at The Geysers (California, USA).

In the *Tuscany–Latium region* (Italy), several geothermal fields have been explored since 1970, including Larderello–Travale and Monte Amiata in the northern region, and Torre Alfina, Latera and Cesano in the southern region. Seismicity of moderate magnitude (up to M 3.2) has been observed around (re)injection wells at the Larderello–Travale field [20] but the relationship with geothermal operations was difficult to identify due to the high natural seismicity in the region. Several events associated with fluid injection in the fractured carbonate rocks were observed at Latera in 1980, including 24 events of magnitude M 1.5–2.0. An event of magnitude M 2.9 occurred during circulation operations in 1984 [10,21]. At Torre Alfina, a sequence of injection tests was performed in 1977 in the fractured limestone. 177 events were recorded within 2 km distance from the well, with the largest event of magnitude M 3.0 being felt by the local population [21,22]. The events ceased soon after the injection was stopped. At Monte Amiata, an event with magnitude 3.5 occurred in 2000 near the Piancastagnaio field [23,24].

The Geysers in Northern California (USA) is the largest producing geothermal field in the world. Although located in a tectonically stressed region, there are no mapped active faults crossing the reservoirs [25]. The Geysers has been in production since the 1960s, with reinjection starting in the 1970s and the substantial addition of two pipelines of treated waste water from local communities, one in 1997 and another in 2002 [26–29]. Seismicity has been monitored for a number of years, creating one of the most comprehensive data sets available. Since the 1980s, more than 750 events have been recorded annually in The Geysers geothermal field. Since 2003, approximately 20 events with $M > 3$ occur every year and 20 $M > 4.0$ events have been recorded, with the maximum, M_w 5.0, occurring in December 2016 [30]. Many studies have shown a strong correlation of spatial and temporal event distributions with injection/production data (e.g., [31–33]; more references in [28]). The concern of the local communities around The Geysers has increased in the last few decades as a consequence of a large number of events.

The pioneering EGS developments at *Fenton Hill* in the USA [34,35], *Rosemanowes* in the UK [36–38] and *Hijiori* in Japan [39] did not produce events large enough to alarm the local population. The maximum magnitudes recorded during stimulation at these sites were, respectively, 1.3, 2.0 and 0.6. However, three months after a long-term circulation test at Hijiori (2000–2002), a M 2.4 event occurred at the edge of the stimulated reservoir [40].

At *Fjällbacka* in Sweden [41,42] the maximum recorded magnitude during stimulation in 1986 was −0.2. A larger event was felt by the project employees during a circulation experiment in 1989, but its magnitude was not recorded. No complaints were received from the local population [10]. It should be noted that the project involved a shallow reservoir at 500 m depth.

Attempts to develop a hot dry rock geothermal production site at *Ogachi* site, northern Japan, at a relatively shallow depth of 1000 m in 1991 were accompanied by more than 1500 seismic events [43]. Except for the maximum event magnitude M 2, all other events had small magnitudes below −1. The hypocenters of observed seismic events were interpreted to be distributed along a natural fault.

6.2.2 After 2000

One of the largest seismic events associated with geothermal energy development occurred in the *Salton* Sea, a region of high geothermal and seismic potential in southern California. The Salton Sea geothermal field has been active since 1982, with multiple injection/production boreholes circulating water in an altered hot sedimentary formation at a relatively shallow depth of 2 km. A large earthquake swarm with a maximum magnitude of M_w 5.1 occurred in 2005 [44]. The earthquake was interpreted to be linked with slip on the largest structural feature in the field, a 5–10 km long fault connecting the shallow geothermal reservoir to deep ones. Indeed, the seismic swarm has migrated 10 km along the fault. As will be discussed in the following, rupture of such a long fault is, in theory, capable of inducing earthquakes as large as M6.

The EGS operations at *Landau* and some 5 km farther at Insheim, south-western Germany, have induced several earthquakes felt by residents. The earthquakes occurred close to populated areas and raised public concerns about geothermal energy exploitation [45,46]. Although the hydraulic stimulation at Landau was almost seismically quiet, the ensuing fluid circulation has promoted seismic activity. In 2009, three years after the project's onset, the largest event of *M* 2.7 took place just after ceasing the operation for maintenance practices [10]. The later stimulation of the reservoir at Insheim in 2010 has been associated with a felt earthquake of magnitude *M* 2.4. Since then, the fluid circulation phase has also witnessed several seismic events comparable in size to those of the stimulation period [45]. Most of the clusters were oriented along a basement fault, providing fluid connectivity across the doublet. Interestingly, the timing of seismic events was closely related to the injected flow rates and increased repeatedly after shut-ins, which confirms the important role of shut-in in induced seismicity [46].

At *Paralana*, Australia, four larger events with *M* 2.7–2.9 occurred during a massive stimulation in 2011. As shown by Albaric et al. [47], the hypocenters of the first three of these events were located at the periphery of the extent of the seismic cloud at that time, whereas the fourth occurred closer to the well, in a zone that had experienced activity earlier during the stimulation process.

Intensive induced seismic activities have been reported at *Hellisheidi*, Iceland [48,49]. Seven deep boreholes have been drilled in the faulted basaltic crust to re-inject the power plant effluent for environmental and pressure maintenance purposes. Earthquakes have repeatedly occurred during the drilling phase due to severe circulation loss into the faults and the subsequent disposal period. More than 4,600 events were located in less than one year (2011–2012), of which 150 were larger than magnitude *M* 2 with the largest event of *M* 4. The injection site is located at plate boundaries with substantial natural seismicity. Although the local population is habituated to small natural earthquakes, the long-lasting injection-induced ground shaking disturbed and frightened them [49]. Thus, this case was of critical importance to the Icelandic geothermal industry, not only due to its large magnitude but also to warn the public, project owners and operators of the challenge of induced earthquakes.

At *Berlin*, El Salvador [50], an event of *M* 4.4 occurred two weeks after the injection shut-in at approximately 2 km from the injection well. Kwiatek et al. [51] found that the seismic activity induced by the injection did not reach distances larger than 500 m from the injection point, and they did not observe migration of seismic activity. However, the injection operations may have changed the dominant fluid migration pathways and indirectly caused the distant event.

More recently, the attempt to develop systems at 4.5–5.0 km depth at Soultz (France), Cooper Basin (Australia), Basel (Switzerland) and Pohang (South Korea) produced events large enough to raise public concerns or cause structural damage, which posed academic and industrial challenges.

The *Cooper Basin* EGS (Australia) comprises two doublets at the Habanero geothermal field [52–54] and one well in the Jolokia field [55]. The six wells have been drilled into the granite at a depth between 3.5 and 4.5 km since 2002, when the exploration started. More than 60,000 events with magnitudes ranging from *M* −1.6 to 3.7 were detected during three hydraulic stimulations: two in Habanero 1 and one in Habanero 4. The maximum recorded magnitudes were *M* 3.7, *M* 3.0 and *M* 3.0, and the microseismic cloud revealed a large thrust fault structure that was well-oriented in the regional stress field and was reactivated by the injection [56]. Stimulation of the Jolokia well, on the other hand, resulted in minor seismic activity ($M < 1.6$) even at extremely high fluid pressures (~120 MPa), where ~200 events with *M* smaller than 1 were recorded and the largest earthquake of *M* 1.6 occurred 127 days after injection ceased. Hypocenters are located around the injection well, suggesting that they occurred on fractures poorly oriented for slip in the regional stress field [19,55]. Despite the major technological advances achieved in the Cooper Basin project, the project was decommissioned in 2016.

The *Soultz-sous-Forêts* project (France) has been developed since 1987 with the drilling of the first well, GPK-1, to explore the granitic basement at about 2 km depth. Subsequently, the well was deepened, and a second well, GPK-2, was drilled, creating a doublet system at 3.0–4.0 km depth in 1992–1997. Between 1998 and 2009, a triplet at 4.5–5.0 km depth was developed by deepening GPK-2 and drilling two additional deviated wells, GPK-3 and GPK-4 [57–59]. Long-term circulation of the deep system with power production commenced in 2010 [60]. All wells were subjected to massive hydraulic injections, with pressures reaching the minimum stress value [61]. While the stimulation and circulation of the doublet at 3.0–4.0 km depth did not produce seismic events larger than *M* 2 [62], the stimulation of the deeper reservoir provoked larger seismicity, although the injected volumes and the stress alteration were comparable. In 2000, the stimulation of well GPK-2 was performed by means of massive water injection for seven days. Some 700 events with magnitudes between *M* 1.0 and *M* 2.5 occurred during injection, but a magnitude *M* 2.6 event was recorded seven days after shut-in during a small-volume injection [28]. Successively, during the stimulation of well GPK-3 in 2003, about 30 microseismic events with $M > 2$ occurred, the greatest with a magnitude of 2.9 two days after the shut-in, despite an attempt to avoid this by a stepwise injection rate reduction [63]. These incidents caused great concern among the local inhabitants, affecting the credibility of the project. The stimulation of the third deep well, GPK4, was performed in September 2004 and February 2005 [64]. Some 128 events with magnitudes between *M* 1.0 and 2.7 were recorded during injection, but none larger than *M* 2.0 occurred during shut-in. Circulation tests have been successfully performed in the triplet in 2005 and in the doublet in 2008, inducing 32 seismic events larger than 1.2 (max 2.3) and 4 events larger than *M* 1.3, respectively. The plant was commissioned in 2008 and is currently operational, producing 1.7 MWe gross power.

The *Basel* EGS project (Switzerland) is probably one of the most controversial cases. A borehole was drilled to a depth of 5 km in the crystalline basement within the city of Basel, with the double purpose of generating power and district heating for the city of Basel. The region is characterized by moderate seismicity, although the city underwent a destructive earthquake with a magnitude of 6.7 in 1356. Stimulation started in December 2006, with a planned 21-day injection. Seismicity began at injection pressures of a few MPa, indicating a critically stressed condition. High rates of microseismic activity built up during the first six days of fluid injection, with more than 10,500 events recorded and their magnitude increasing with the pressure [65]. After an event of magnitude 2.6 on the fifth day of injection, it was decided to reduce the injection for 5 hours before shutting in. Two events of *M* 2.7 and *M* 3.4 occurred during shut-in and preparations for bleeding off the well to hydrostatic conditions [65–68]. Over the following months, seismic activity declined rapidly, although three aftershocks of $M > 3$ and thousands of smaller events were recorded, leading to the suspension and eventual cancellation of the project in 2009. The hypocenter location at Basel lies between 4 and 5 km depth and forms an elongated, steeply dipping structure of 1.2 km diameter trending N155 ± 5°E, with an offset branch striking N100 ± 10°E. Overall seismicity migrates away from the borehole during the injection, supporting the conjecture that the diffusion of overpressure drives the seismicity. After shut-in and bleed-off, seismic activity is mostly located at the periphery of the hypocenter cloud [67]. Among the 28 strongest events, most exhibit fault plane solutions of pure strike-slip mechanisms [66], which is in agreement with the regional stress field. However, the strike of the nodal planes of most of the events is oblique to the overall orientation of the seismic cloud. This suggests that the internal structure of the stimulated rock volume is composed of a complex network of individual fault segments arranged "en échelon" with alternating NS and EW orientations along the presumed cataclastic fracture zone [65]. Despite the diverse attempts to explain or model the seismicity observed at Basel (see Section 6.5), the occurrence of such a large event and, in particular, magnitude jump in the post-injection stage remained largely unresolved, which reduced confidence in geothermal energy technology.

More recently, in 2017, another case of post-injection seismicity at *Pohang* EGS (South Korea) caused extensive damage and posed grand scientific challenges. A series of five massive injections have been performed to stimulate the two wells drilled into the crystalline basement to a depth of 4.3 km. These stimulations caused seismic activity reaching a magnitude of up to 3.1 in April 2017 (we refer to the local magnitudes recorded by the Korea Meteorological Administration, KMA). Two months after the completion of the fifth stimulation, a magnitude 5.4 earthquake occurred at 4.3 km depth [69–73]. The earthquake caused extensive injuries to local inhabitants and significant damage to structures. As a consequence, the project was suspended, and a commission was appointed by the government to investigate the case. Results from this study concluded that the M 5.4 earthquake had been triggered by the hydraulic stimulations, based on the spatio-temporal correlation between hydraulic stimulation and observed seismicity [74]. The Pohang EGS project was, therefore, officially terminated in April 2019. Many research efforts have been made since then to identify the causal mechanisms of such unprecedented, long-time delayed, high-magnitude seismic activity (see Section 6.5). The Pohang event is highly unprecedented in EGS-induced seismicity due to two aspects. First is the obvious outlier of injection volume-based magnitude prediction (see Section 6.5.2.1). The total injected volume was 12,798 m3 with a bleed-off of 5,841 m3. Thousand folds of injection volume would have to be needed to cause magnitude 5.4 based on estimation by injection volume [75]. The second is the largest magnitude jump during shut-in in EGS operations. The largest magnitude before M 5.4 event was M 3.1 in April, and magnitudes larger than 2.0 did not occur in the fourth and fifth stimulations in August and September, respectively. This magnitude jump from 3.1 to 5.4 is the first of its kind in EGS history and poses a great challenge to managing the induced seismicity.

A review of observed cases has allowed for highlighting of some general trends. For example, the observed magnitude of induced earthquakes positively correlates both with depth and rock strength [10]. Large earthquakes occur at depths larger than 4 km, and they tend to nucleate in crystalline rocks rather than in sedimentary rocks (Table 6.1 and Figure 6.1), where failure tends to be ductile because of their moderate strength. However, the correlation with depth is not always satisfied [10]. The lack of spatial and temporal correspondence between the injection-induced pressure increase and the seismicity occurrence and magnitude reveals the complexity of the problem, which deserves a deeper analysis.

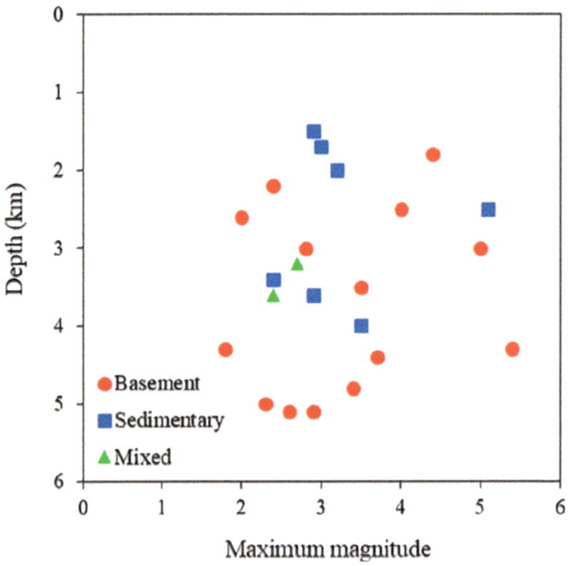

Figure 6.1 Maximum magnitude observed in EGS systems as a function of depth and rock type. Basement refers to both crystalline and metamorphic rocks. Mixed refers to cases where the depth of occurred seismicity is not well constrained (e.g., Insheim and Landau, see Table 6.1).

6.3 PHYSICS OF EARTHQUAKES

6.3.1 Earthquake quantification

Earthquakes are failure processes occurring on fault planes. When stress builds up such that failure conditions are reached, inelastic shear slip occurs, which releases the energy stored in the rock in the form of an earthquake. Following the slip activation, shear stress drops, and equilibrium is restored. Despite the simplicity of the process, the details are complex, mostly because of the temporal variability of loading, rock strength and stress drop [76].

The size of an earthquake (or how much energy is released from it) depends on how much slip occurs along the fault, how much stress there is on the fault before slipping, how fast it fails and over how large an area it occurs [77]. The most established method for measuring earthquakes is to measure the elastic wave field generated by the sudden slip on a fault plane. According to the elastic dislocation theory, the seismic moment M_0 can be calculated as [78]

$$M_0 = G \cdot D \cdot S, \tag{6.1}$$

where G is the rigidity of the material surrounding the fault (the elastic shear modulus), D is the slip displacement offset (difference between the displacements of the two sides of the fault) and S is rupture region surface area.

Despite the accuracy of seismic moments, magnitude scales were historically used to quantify the size of an earthquake. Most magnitude scales were defined by the observed amplitude of seismic waves at the surface, with some corrections for attenuation with distance from the source. The Richter scale for local magnitude, M_L [79], is widely adopted, although it underestimates the magnitude of earthquakes at greater depths, distances, or magnitudes, a problem called saturation. Recently, the standard practice is to define the magnitude of the seismic event, M_w, with the seismic moment, M_0, through [80–82]

$$M_w = \frac{2}{3}\log M_0 - 6.07 \ \left[M_0 \text{ in Nm}\right], \tag{6.2}$$

which provide more accurate measures. The relationship between M_L and M_w has been discussed by seismologists for years, as they are measured by different techniques and a simple theoretical relation between them does not appear to exist. Numerous studies have been carried out to obtain empirical relationships between them. A theoretical ratio has been proposed by [83], such that M_w / M_L varies within the range from 2/3 to 4/3.

The average shear stress drop, $\Delta\overline{\tau}$, consequent to shear activation is [76]

$$\overline{\Delta\tau} \approx \frac{C{\cdot}G{\cdot}\bar{D}}{L}, \tag{6.3}$$

where $\bar{D}$ is the average slip over the faulted region, L is a characteristic rupture length, often defined as $\sqrt{S}$, and C is a geometric constant of unitary order.

Substituting Eq. 6.3 into Eq. 6.1, and assuming $L = \sqrt{S}$, we obtain

$$\mathrm{M}_0 = \frac{\overline{\Delta\tau}}{C}\, S^{3/2}\,. \tag{6.4}$$

According to this expression, if $\overline{\Delta\tau}$ is constant, we find the scaling relationship $M_0 \propto S^{3/2}$, which has also been empirically estimated from observations of earthquakes [76,84].

The scaling expressed by Eqs. 6.1–6.4 allows for making some considerations. Damaging earthquakes (usually greater than M 4 or M 5) require the surfaces to slip over relatively large distances (kilometer-scale), which tends to be possible only on developed or active fault systems. Large earthquakes rarely occur on small faults, and it is also difficult to create a large new fault because there is usually a pre-existing fault that will slip first. Most geothermal fields with observed seismicity are located in regions with tectonic activity (see Section 6.2), which are more prone to seismicity than more stable areas. Moreover, it has been shown that, in almost all cases, large earthquakes nucleate at depths of at least 5–10 km [77], where the stored energy is enough to move large volumes and trigger large earthquakes (Figure 6.2).

Despite the simplicity of these considerations, there are many physical and micro-physical processes involved in the occurrence of earthquakes, such as frictional weakening, melting and fault fluid pressurization due to mechanical compression or heating during frictional sliding [76,85,86]. Moreover, the complexity of the fault rheology and fault zone structure may affect the nucleation time. To overcome this complexity, a successful approach is to predict earthquakes by means of a statistical

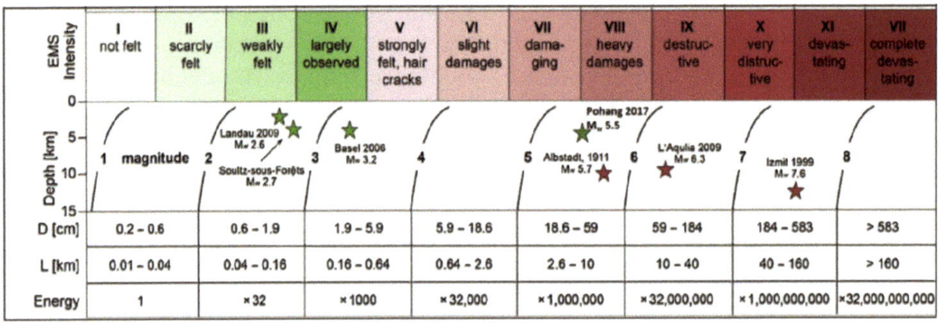

Figure 6.2 Earthquake fault scaling relationship. Rows indicate seismic intensity (*EMS*), depth and earthquake moment magnitude M_w in relation to slip displacement *D*, rupture length *L* and energy of the rupture process. Selected EGS seismicity (green stars) and tectonic earthquakes (red stars) are shown in the depth section. The boundary between intensity IV (dark green) and intensity V (light red) is indicated by a change in color. Adapted from Ref. [18], with permission from Elsevier.

analysis. In statistical seismology, there are two fundamental laws, namely the Omori law, which describes the decay rate of aftershock activity, and the Gutenberg–Richter relation, which describes the frequency-magnitude distribution of earthquakes [87].

After a large earthquake (the mainshock), many smaller earthquakes, called aftershocks, occur near the rupture zone of the earthquake. As expressed by Omori law [88,89], the decay of aftershock activity follows a power law:

$$n(t) = \frac{\psi}{t + c} \tag{6.5}$$

where $n(t)$ is the number of aftershocks larger than a given magnitude per unit time, t is the time after the mainshock occurrence, c and ψ are constants.

The frequency-magnitude relation defined by the Gutenberg–Richter (G-R) relation [90] states quantitatively that small earthquakes are more frequent than large earthquakes. This relation describes the number of earthquakes of a given magnitude expected in a given area. In any area much larger than the rupture area of the largest earthquake considered, the number of earthquakes, $N(M)$, with a magnitude greater than or equal to M is given by the relation

$$\log N\,(M) = a - b \cdot M, \tag{6.6}$$

where a and b are constants to be estimated from the event catalogs. For natural tectonic earthquakes, the value of b is nearly 1, which means that there are statistically ten times more events of magnitude 2 than magnitude 3 for a given time period. However, the *b*-value can vary. Higher *b*-values indicate more small events relative to larger events and vice versa. At Basel, the *b*-value was larger during the injection period (~1.58) than during the post-injection period (~1.15) [91].

The G-R relation is a key tool in probabilistic hazard assessment, as it allows assessing the probability of large earthquakes by means of extrapolation from the rates of small events. Based on the Gutenberg–Richter law of frequency-magnitude distribution and the Omori law of aftershock decay, the seismic hazard can be forecasted based on the previously observed seismicity catalog [92–94].

6.3.2 Rock failure

Earthquakes are activated when failure conditions are met in the rock, which is strictly related to the subsurface stress balance. Subsurface forces are described in terms of three principal stresses acting perpendicular to each other and numbered in order of decreasing magnitude, i.e., $\sigma_1 > \sigma_2 > \sigma_3$. It is common practice to assume they are aligned with the vertical and two horizontal directions. The relative magnitude between the vertical, σ_v, and the two horizontal stresses, σ_H and σ_h, defines the in-situ stress regime (normal faulting if $\sigma_v > \sigma_H > \sigma_h$, strike-slip faulting if $\sigma_H > \sigma_v > \sigma_h$, reverse faulting if $\sigma_H > \sigma_h > \sigma_v$), according to Anderson's classification [95,96] (Figure 6.3). The vertical stress is usually estimated as the lithostatic stress, while the horizontal stresses are estimated using a range of measurements in well bores, including fracture and leak-off tests, caliper measurements of wellbore breakout and imagery of the strike angles of existing fractures [97].

Rock failure may occur in either a tensile or shear mode [98]. Tensile failure is likely to occur along planes normal to the minimum principal stress, σ_3, when the pore fluid pressure, P, equals or exceeds the sum of σ_3 and the rock tensile strength, T_o, as expressed by the failure criterion

$$P = \sigma_3 + T_o. \tag{6.7}$$

This is the basic assumption for hydraulic fracturing operations [99], which open fresh fractures and/or widen and propagate pre-existing fractures.

In contrast, shear slip along newly created or reactivated faults occurs under a state of compressive stress in the rock. It is widely believed that the dominant mechanism in the hydraulic stimulation of a fractured rock mass is the activation of shear

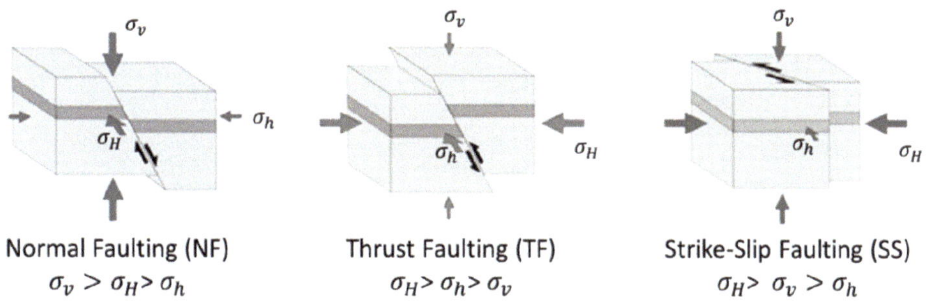

Figure 6.3 Fault regimes according to Anderson's classification.

failure on existing fractures – termed hydroshearing - rather than the creation of new ones [36]. Due to the roughness of the fracture surface, shear sliding is usually accompanied by dilation that increases the fracture aperture, thus the permeability, permanently. Shear failure is triggered along a given orientation when the shear forces overcome the frictional resistance [100], according to the classical Coulomb failure criterion (Figure 6.4)

$$\tau = c + \mu(\sigma_n - \alpha P), \tag{6.8}$$

where τ and σ_n represent the shear stress acting parallel to the orientation and the compressive stress acting normal to the orientation, respectively, c is rock cohesion, μ is the coefficient of internal friction, which is often expressed in terms of the angle of internal friction φ, e.g., $\mu = \tan\varphi$, α represents the Biot-Willis coefficient, with values between 0 and 1. Note that we adopt the sign convention of geomechanics, in which compressive strain and stresses are positive. If the shear stress exceeds the frictional forces opposing slip, there is the possibility for fault reactivation to occur in the form of seismic or aseismic sliding [101].

Geometrical relationships link the stress acting on a plane, τ and σ_n, with the principal stresses in the system, as expressed by the Mohr circle graphical representation, typically presented together with the failure criterion (Eq. 6.8) in the Mohr-Coulomb representation (Figure 6.4). Shear failure occurs when the circle touches the failure envelope, with the contact point representing the orientation of failure. This representation allows for illustrating a number of features relevant to the failure potential. Firstly, failure depends on the initial stress state and on the orientation of the weak plane relative to the orientation of the principal stresses. The initial stress may already be close to failure (critically stressed), depending on the maximum and minimum principal stresses. A larger differential stress $\sigma_1 - \sigma_3$ (large diameter of the Mohr circle) and a smaller mean effective stress $(\sigma_1 + \sigma_3)/2$ correspond to a state of stress closer to

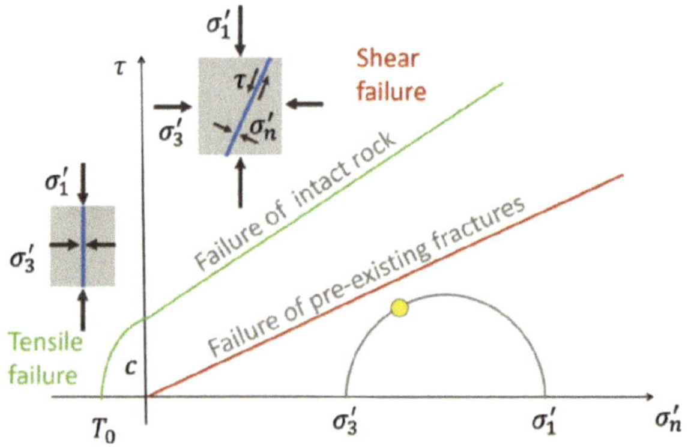

Figure 6.4 Mohr-Coulomb representation: Mohr diagram of shear stress (τ) versus effective normal stress (σ_n') corresponding to the example stress regime, and Coulomb failure envelopes for intact rock (green line) and pre-existing fractures (red line). Failure occurs when the Mohr circle touches the failure envelope. The yellow dot represents the critical orientation for shear failure, i.e., the first point of the circle that touches the red envelope if the circle displaces or enlarges toward the failure envelope.

failure. Reactivation of critically stressed faults may require only small stress changes, whereas reactivation of stable faults requires large stress changes. The most critical orientations for failure are the ones forming angles of $(\pi/4 \pm \varphi/2)^{\circ}$ with the direction of the maximum principal stress, σ_1. These optimally oriented planes are therefore defined by the stress regime (normal faulting NF, strike-slip SS, reverse faulting RF) according to the Andersonian fault classification, as schematically represented in Figure 6.3. Second, failure depends on the rock cohesion c, and the coefficient of internal friction, μ. The more cohesive the rock, the greater the difference in minimum and maximum stress states (the radius of the Mohr circle) required for failure to occur. Similarly, the larger the internal friction coefficient, the farther the failure line will be from the circle of stress state. The conservative assumption of $c = 0$ acknowledges that shear failure is likely to occur along planes of weakness, e.g., faults.

The difference between stress and pore fluid pressure is the effective stress [98],

$$\sigma_n' = \sigma_n - \alpha P, \tag{6.9}$$

reflecting the effects of fluid pressure on the amount of stress that acts on the solid skeleton. While natural earthquakes are caused by an increase in shear stress τ, typically caused by tectonically driven deformations [95,96,102,103], induced seismicity is typically driven or triggered by a reduction in effective normal stress caused by an increase in fluid pressure associated with fluid injection (Eq. 6.8). Pressurizing the pore fluid impacts the effective stresses as both the minimum and maximum compressive principal effective stresses will be reduced by α times the amount of pressure increase in the system, ΔP. The increase in pore pressure has the effect of translating the circle of stress to the left, bringing it closer to the line of failure. The fluid injection may, however, result in further impacts, including poroelastic and thermoelastic effects, which may change the principal stresses by different amounts, as we analyze in the following section.

6.4 EARTHQUAKE TRIGGERING MECHANISMS IN SUBSURFACE INJECTION ACTIVITIES

A number of physical processes occur during EGS operations that have the potential to trigger seismic activity [104]. They are strictly related to each other, and their combination is in general denoted with the expression coupled *thermo-hydro-mechano-chemical processes* (THMC) or just *coupled processes*. The most significant of these processes are detailed in the following. While extraction or production of geothermal fluid can also trigger earthquakes [5,105], we mainly focus on the injection related processes. Additional mechanisms that we do not analyze here are fluid pressurization due to mechanical compression, heating during frictional sliding and frictional weakening [85,86,106]. Likewise, we do not discuss other mechanisms that may induce seismicity in fluid production activities.

6.4.1 Pressure-driven stress variation

Fluid pressure buildup is considered as the main mechanism inducing seismicity because it causes a reduction of the effective stresses and, therefore, of the frictional resistance (see Section 6.3.2). Consequent to fluid injection, pore fluid pressure increases, and this overpressure diffuses away from the injection well as a function of time. The process is governed by the fluid flow equation [107]

$$\phi\beta\frac{\partial \Delta P}{\partial t}+\frac{\partial}{\partial t}(\nabla\cdot\mathbf{u})=\nabla\cdot\left(\frac{k}{\eta}\nabla(\Delta P)\right), \tag{6.10}$$

where ΔP is the pressure, t is the time, ϕ is the porosity, β is the fluid compressibility, $\mathbf{u}$ is the displacement vector with $\nabla\cdot\mathbf{u}$ representing the volumetric deformation, k is the intrinsic permeability, η is the fluid viscosity and ∇ refers to the space coordinate field. The equation is usually written by assuming that the volumetric deformation is proportional to P through rock compressibility. In this case and in the hypothesis of a homogeneous reservoir, the overpressure reduces with the logarithm of the distance from the well, and the pressurized volume, i.e., the volume affected by pressure buildup, increases with the square root of time. However, the presence of heterogeneities and fractures affects the response to pressure because they may act as either preferential paths or barriers for flow. Geothermal wells often intercept fault zones with a larger transmissivity than the surrounding rocks, resulting in a non-axisymmetric pressure distribution around the well. When the injection is terminated, overpressure drops at the injection well, but it still propagates at the periphery of the pressurized region.

Failure may be induced by pore pressure increases, through direct variation of the effective stress (Eq. 6.9), but also by poroelastic strains. The elastic deformations consequent to pressure variation, in the absence of body and surface forces, are expressed by the mechanical equilibrium [98]

$$\nabla\cdot(G\nabla\mathbf{u})+\nabla\cdot\left[G(\nabla\mathbf{u})^T+\lambda(\nabla\cdot\mathbf{u})\mathbf{I}\right]=-\nabla(\alpha\Delta P), \tag{6.11}$$

where λ is the elastic Lamé modulus and $\mathbf{I}$ is the identity matrix, while the total stress tensor variation is expressed by [108]

$$\Delta\sigma=G\left[\nabla\mathbf{u}+(\nabla\mathbf{u})^T\right]+\lambda(\nabla\cdot\mathbf{u})\mathbf{I}+\alpha\Delta P\mathbf{I}. \tag{6.12}$$

Eqs. 6.10 and 6.11 are strictly coupled and constitute the hydro-mechanical (HM) problem. Variations of the pressure field provoke elastic deformations and stress perturbations, while deformations, in turn, affect the pressure field because they change the available storage volume (note that $\nabla\cdot\mathbf{u}$ corresponds to a change in porosity). The effective stress variation resulting from this coupled behavior is different from the one predicted by the direct application of the overpressure, ΔP, to Eq. 6.9, because the total stress does not remain constant, both inside and outside the pressurized region (e.g., [109–111]). Indeed, poroelastic mechanisms can cause stress changes even at distances greater than the region of pressure perturbation, depending on the system geometry and boundary conditions [105,112]. In the case of a spherical volume, the poroelastic stress is isotropic, while if the pressurized volume is elongated, the induced stress is larger in the direction of the elongation (Figure 6.5). The presence of fault zone structures, or heterogeneity in general, alters the hydraulic and, consequently, the geomechanical response with respect to homogeneous media.

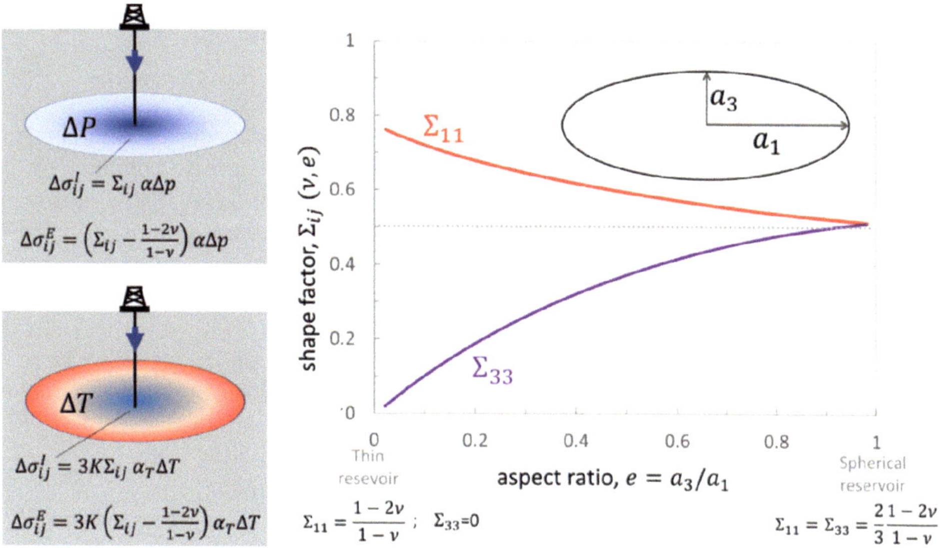

Figure 6.5 Schematic illustration of poroelastic and thermoelastic total stress variations induced inside (superscript *I*) and outside (superscript *E*) of a pressurized and a cooled ellipsoidal volume, respectively. Both poroelastic and thermoelastic stresses depend on the shape factor Σ_{ij}, which is a function of the volume aspect ratio, e, and of Poisson's coefficient, ν, as illustrated in the plot for the case of ν = 0.2 (formal expressions in Ref. [120]). Note that, differently from the poroelastic stress, the thermoelastic stress increases with the rock stiffness, *K*.

In fact, the pressurized region is often ellipsoidal rather than spherical, with the consequence that stresses induced by these processes are heterogeneous and anisotropic [5]. In particular, the total compressive stress parallel to the flow direction significantly increases because of differential deformations, which implies that the reduction of the effective stress parallel to the flow will be smaller than the reduction of the stress normal to the flow direction [113–115]. This may encourage or inhibit seismicity in a fracture depending on its location and orientation and on the in-situ stress regime [116,117] (see also Section 6.5.2).

These coupled HM interactions can be direct, as explained above, but also indirect [118]. In fact, rock deformation corresponds to a variation of the porous volume or of the fracture aperture, which impacts the hydraulic transmissivity. At Soultz, the net transmissivity increased by 15 times as a result of the fluid-injection operations [59,119].

6.4.2 Thermal effects

Fluid injected into the reservoirs is usually at a colder temperature than the deep rock. The temperature contrast provokes significant deformation of the reservoir, which affects the local stress field. Moreover, the thermal contraction has effects on the flow process, which, in turn, affects the advective heat and solute transport.

Heat transport is governed by the energy balance [107]

$$\frac{\partial \Delta T}{\partial t} + q\frac{C_w}{C}\nabla \Delta T = \frac{\zeta}{C}\nabla^2 \Delta T, \tag{6.13}$$

where q is the fluid specific flux, C_w and C are the fluid and reservoir heat capacity, and ζ is the thermal conductivity. In the case of EGS, the area affected by temperature variation is smaller than the pressurized region at any given time [104]. Considering typical values [98] and a continuous injection for the time t, the pressure propagation distance is in the order of $\sim 10^{-2}\sqrt{t}$ (m, with t expressed in s) in granitic rocks, whereas the thermal propagation, governed by conduction due to the low permeability of granite, is in the order of $\sim 10^{-3}\sqrt{t}$.

The thermoelastic response is analogous to the poroelastic one, as illustrated by many authors [121–123]. The elastic strains in response to the cooling down of the rock are indeed expressed by

$$\nabla \cdot (G\nabla \mathbf{u}) + \nabla \cdot \left[G(\nabla \mathbf{u})^T + \lambda(\nabla \cdot \mathbf{u})\mathbf{I}\right] = -(2G + 3\lambda)\alpha_T \nabla(\Delta T), \tag{6.14}$$

where α_T is the linear thermal expansion coefficient. Likewise, thermal stresses are expressed by

$$\Delta\sigma = G\left[\nabla \mathbf{u} + (\nabla \mathbf{u})^T\right] + \lambda(\nabla \cdot \mathbf{u})\mathbf{I} + (2G + 3\lambda)\alpha_T \Delta T\mathbf{I}, \tag{6.15}$$

and they arise depending on the geometry of the cooled region, on the presence of lateral constraints, or when differential deformations occur (Figure 6.5). Similarly to the poroelastic stress, thermoelastic stresses are proportional to the magnitude of the temperature drop ΔT and they depend on the shape of the cooled region. However, differently from poroelastic behavior, the magnitude of the thermal stress is proportional to the rock stiffness, so it may be significant in sedimentary or crystalline rocks [124,125].

6.4.3 Geochemical effects

The reservoir rock minerals and the resident pore fluid have reached chemical equilibrium over geological time scales. The injection of cold, non-native fluid for EGS development disturbs the initial chemical equilibrium. As a result, a set of geochemical reactions, mainly in terms of mineral dissolution and precipitation, can be expected. Generally, if the injected fluid is undersaturated with respect to host rock minerals, mineral dissolution is likely to occur, while mineral precipitation occurs when the fluid flowing through the rock becomes supersaturated. In the meantime, cooling effects raised by cold water injection change mineral solubilities, promoting their dissolution or precipitation [126]. Interestingly, cold water injection favors calcite and anhydrate dissolution in the region close to the injection borehole because their solubility rises with the decreasing temperature, a phenomenon widely known as retrograde solubility [127]. On the contrary, the solubility of silicates, sulfides, fluorite, barite and halite decreases within the cooled region, i.e., prograde solubility [128,129], leading to their precipitation.

The rate of chemical reactions expressed as dissolved/precipitated mineral mass per unit volume of rock and unit time significantly differs among different minerals and varies mainly with temperature and pH, among others [130]. The content of fast-reacting minerals primarily determines the fate of geochemical reactions and their influence on the reservoir response to fluid injection. In commonly encountered granitic geothermal reservoirs, dissolution of calcite and precipitation of amorphous or poorly crystalline silica take place much faster and more frequently than for other minerals, particularly feldspar and quartz [131–133]. As an example, calcite dissolution at the 160°C Soultz-sous-Forêts reservoir has been estimated to occur at a maximum rate of 3 × 10−4 mol m−3 s−1, while quartz precipitation rates have hardly approached 2 × 10−8 mol m−3 s−1 [134].

Silicates (SiO_2) are hydroxide soluble, meaning that a higher pH (alkaline conditions) favors their solubility in water, whose reaction is as follows:

$$SiO_{2\,(am)} + 2H_2O \leftrightarrow Si(OH)_{4\,(aq)}, \tag{6.16}$$

where $Si(OH)_4$ is silicic acid, which can remain monomeric at neutral or acidic pHs or undergo ionization and produce silicate anions in alkaline conditions [135]. As soon as the formation brine meets supersaturation conditions with respect to acidic acid, probably during injecting silica-rich fluid or cooling, amorphous silica precipitates [129].

On the contrary, carbonates are acid-soluble, featuring higher solubility and faster reactions at low pHs (acidic conditions). The dissolution and precipitation reactions of calcite ($CaCO_3$) in water are described as [136]:

$$CaCO_{3\,(s)} + H_2O \leftrightarrow Ca^{2+} + HCO_3^- + OH^-. \tag{6.17}$$

As a proton-consuming reaction, calcite dissolution also depends on the partial pressure of CO_2, producing carbonic acid (H_2CO_3) in water:

$$CO_{2\,(aq)} + H_2O \leftrightarrow H_2CO_3 \leftrightarrow HCO_3^- + H^+, \tag{6.18}$$

which reacts with calcite as:

$$CaCO_{3\,(s)} + H^+ \leftrightarrow Ca^{2+} + HCO_3^-. \tag{6.19}$$

CO_2 may be naturally present in geothermal reservoirs or injected as the circulating fluid to enhance heat extraction efficiency and store CO2 underground [137]. While calcite dissolution occurs under acidic conditions, the pH increase and retrograde solubility of calcite likely cause it to precipitate.

Geochemical reactions alter the rock's pore structure and, thus, its porosity. The porosity increases with the net dissolution and removal of minerals, whereas it decreases with mineral precipitation. These phenomena are more pronounced within fractures, controlling flow and heat extraction efficiency in naturally or hydraulically fractured geothermal systems [126,138]. The chemical effects may significantly change the fracture aperture. Demonstrably, numerical simulations and post-injection field observations have pointed to drastic permeability changes in Coso and Soultz-sous-Forêts geothermal fields associated with a combination of calcite dissolution and silica precipitation in the region close to the injector and calcite precipitation adjacent to the producer [131,134,139]. Chemical reactions may become exceedingly important in intersecting fractures. Fluid mixing where fractures of different origins join together or splitting where fractures branch may locally alter the fluid velocity field and chemistry, which, in turn, may either promote or hinder chemical reactions [140]. In addition to mineral reactions on free fracture surfaces, cold water injection may provoke mineral dissolution when contacting load-bearing asperities in fractures. Dissolved species subsequently diffuse through the grain boundary fluid film and may precipitate elsewhere within the fracture. This geochemical process is widely referred to as pressure solution [141,142]. It contributes to further compaction of fractures and, thus, loss of permeability in the long term [143–145].

Overall, chemically-induced alterations in fracture conductivity take place in a heterogeneous and non-monotonic manner, corresponding to the spatial and temporal evolution of mineral solubility and reaction rates as first-order functions of temperature, pressure and pH [128,146]. Therefore, thermal and flow paths are likely to change as a result of chemical reactions, redistributing temperature and pore-pressure profiles and, in turn, the state of stresses, whose effects on fault reactivation have been discussed in Sections 6.4.1 and 6.4.2. Furthermore, the dissolution of asperities and altering the mineralogical composition of fracture surfaces may result in friction degradation of the fracture, making it more prone to slip [147,148]. However, coupled reactive transport and geomechanics simulations of fluid circulation in sedimentary or crystalline geothermal reservoirs imply that geochemical effects operate more slowly than hydraulic and thermal processes (relatively longer characteristic times of geochemical interactions [131,146,149,150]). Accordingly, the reservoir response in terms of permeability and stress evolutions is principally dominated by hydraulic and thermal processes in the short (hours to a few days) and intermediate (days to a few months) terms, respectively, whereas geochemical effects onset after several months mainly by dissolution/precipitation of calcite and become increasingly important by the precipitation of amorphous silica after several years. Notwithstanding their slow nature, geochemical processes may notably influence the long-term reservoir life primarily by changing the reservoir injectivity and propensity to induce seismicity [131,147,149].

6.4.4 Stress transfer

The activation of a seismic or microseismic event provokes a stress variation around the slipped surface, which may in turn promote failure on other faults. Specifically, there is a reduction of shear stress around the slipped surface (stress drop) and anti-symmetric variations of both the stresses perpendicular and parallel to the fault at the tips of the slipped surface (Figure 6.6).

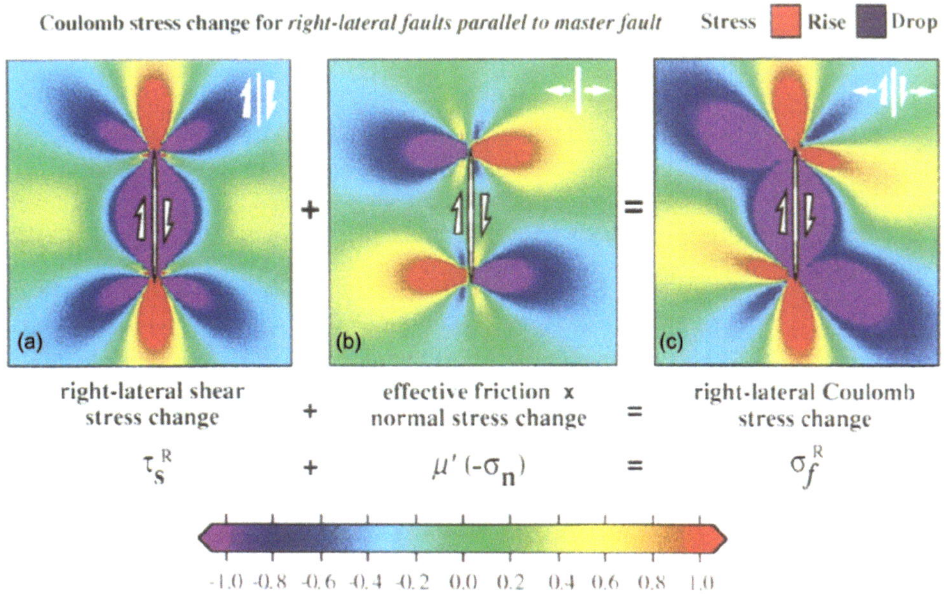

Figure 6.6 Map view of the stress variation in an elastic space consequent to right-lateral strike slip. Shear (a), normal (b) and Coulomb stress (c) variations for a right-lateral fault parallel to the slipping fault. The cohesion is assumed to be zero. Reprinted from Ref. [151], with permission from the Seismological Society of America under STM permissions.

This stress perturbation is permanent because slip is irreversible, and it modifies the stability conditions, as usually quantified by means of the Coulomb Stress change (or Coulomb Failure Stress change), which is directly derived from Eq. 6.8 as equal to $\tau - (c + \mu\sigma_n')$.

Shear stress transfer governs the aftershock events in tectonic seismicity [151], and it has been shown to be relevant in induced seismicity [114,152]. The magnitude of the slip-induced shear and normal stresses scales with the rock stiffness and amount of slip, whereas it is inversely proportional to the size of the slipping surface [153] (recall Eq. 6.3). The size of the area affected by slip-induced stress is instead proportional to the dimension of the slipping surface.

The variation of the stress field does not depend on the in-situ initial stress field, although the occurrence of slip does. The static stress transfer produces a rotation of the stress tensor, which, contrary to the variation of the stress field, does depend on the initial stress state. Specifically, it depends on the ratio of the shear stress variation to the initial in-situ shear stress [151]. Therefore, the smaller the initial stresses, the greater the rotation for a given shear stress variation. The stress tensor rotation changes the orientation of the faults that are critically oriented to undergo shear failure.

Notice that static stress transfer is also provoked by aseismic shear slip, e.g., shear-slip failure that does not produce an earthquake [101], with the potential to induce seismicity [154,155], as further discussed in Section 6.5.2.

6.5 MODELING OF INDUCED SEISMICITY

The risk of inducing high levels of seismicity during reservoir stimulation should ideally be predicted before the expensive drilling phase has started. However, there are no consolidated or reliable methods for forecasting induced seismicity hazards due to injection to date. The commonly used method in EGS is the so-called traffic light system [50,156], a reactive control method consisting of planning a series of adaptative actions to implement whenever one or more indicators exceed the predefined thresholds. Typical indicators are peak ground velocity, event magnitude, the cumulative number of seismic events and public perception. Actions like decreasing the injection rate, decreasing the well head pressure, stopping the injection or relieving the well head pressure (bleed-off) are planned according to the green, amber or red lights [17]. An intrinsic drawback of the approach is that the adaptation operations are often implemented too late. Moreover, defining adequate indicators, thresholds and actions is difficult.

To avoid harmful earthquakes by taking the correct decision at the right time, a challenging approach includes real-time seismicity modeling during injection to build the so-called Adaptive Traffic Light System. This method relies on the ability to model time-dependent seismic hazards. Therefore, research efforts over the last three decades have been oriented to advance the understanding of the governing processes and factors and to improve the modeling of such processes, with the ultimate goal of seismicity control and mitigation [14,157–159]. Different approaches have been proposed that can be classified into three categories (see also the review by Gaucher et al. [160]).

6.5.1 Statistical modeling

Based on the statistical seismology concepts described in Section 6.3.1, the statistical approach is aimed at describing and forecasting the probability of an earthquake's spatio-temporal occurrence and magnitude. The catalog of the observed events is analyzed to calibrate statistical modeling based on the G-R frequency-magnitude relation (Eq. 6.6) and the Omori law of aftershock decay (Eq. 6.5). These models allow for forecasting seismic hazards in a given time window based on previous observations.

By employing these methods, Wenzel et al. [161] estimate earthquake probabilities and expected magnitudes for post-injection seismicity, assuming a constant b-value. Barth et al. [162] show that time-decreasing b-values are necessary to explain the observed increase in seismic hazard after shut-in. By using Omori's law and the G-R distribution with a reduced b-value in the post-injection phase, they demonstrated that the probability of exceeding a certain maximum magnitude still increases after shut-in. This kind of approach was also tested on the Basel sequence by Bachmann et al. [91]. They find that the b-value significantly decreases from the co-injection to the post-injection period, which indicates an increase in the seismic hazard. Regardless of the location, they attempt to reproduce the earthquake's temporal occurrence and magnitude by using different statistical models. They find that the incorporation of background seismicity associated with the injection rate provides better predictions of the observed data.

The advantage of statistical models is that they are simple, easy to implement and well established in statistical seismology. What makes such models simple, however, also constitutes their weakness because the models are independent of the physics of the triggering mechanisms. Furthermore, they require seismicity observations spanning a wide range of magnitudes in a given time window to calibrate their use. Therefore, they have limited predicting capabilities for longer time periods, which means that they cannot be used to predict the risk of induced seismicity prior to initiating injection and in the post-injection period or to establish alternative injection strategies. For all these reasons, statistical analysis should be considered only as a starting point or a reference for the assessment of the time-dependent seismic hazard [91].

Shapiro et al. [163] and Mena et al. [164] improved the performance of statistical modeling by introducing the injection volume as information. Shapiro et al. [163] suggest the estimation of a site-specific seismogenic index based on the observed rate of induced events and the fluid-injection rate. This index is independent of injection parameters and defines the tectonic tendency toward seismicity. Combining this index with injection parameters makes it possible to estimate the probability that a given number of events occur during the injection. Since the event probability is assumed to be proportional to the pore-pressure perturbation, the cumulative number of induced events is proportional to the injected volume, under the assumption of constant or increasing pore-pressure. The seismogenic index and the injected volume are introduced into the Gutenberg–Richter Law to predict the number of events with a magnitude larger than a given one. The model was also applied by Mena et al. [164], in combination with a statistical aftershock model. The models proposed by Shapiro et al. [163] and Mena et al. [164] highlight the importance of considering physical processes in the assessment of induced seismicity, which led to the creation of hybrid physical-statistical models, as we discuss in Section 6.5.3.

6.5.2 Physics-based models

Physics-based models consider the physical processes leading to seismic activity according to two different approaches. In the analytical approach, the dependency of the seismic response on physical processes is sought by looking at the correlation between observed seismic activity and one or more controlling parameters, which depend on simple governing equations that can be solved analytically. In the numerical approach, one or more physical processes are simulated in the context of geomechanical modeling to reproduce the seismic activity.

6.5.2.1 Analytical approach

Analytical modeling is aimed at finding the factors that govern seismicity by considering simple physical processes and looking at the correlations between seismic activity and key parameters. A popular approach suggests that induced seismicity scales with the volume of injected fluid, ΔV. McGarr [75] updated the originally proposed relation between the injected volume and the cumulative seismic moment [4] by suggesting that the maximum seismic moment that can be released is limited by a quantity proportional to the injected volume. The scaling is based on simple considerations of proportionality between maximum seismic moment, induced shear stress, fluid-induced strains and overpressure. This upper threshold is validated against a collection of case histories of injection-induced seismicity, including, but not limited to, EGS. This relationship has been found to be valid for numerous injection experiments, and the maximum magnitude of earthquakes is observed to be much lower than the volume-based estimation [165]. However, cases where earthquakes with magnitudes larger than suggested by McGarr's upper limit have been observed, in particular with post-injection seismicity (Figure 6.7).

Overcoming the simplified assumption of seismicity occurring in the stimulated volume, Galis et al. [166] consider theoretical estimates of the rupture size induced by localized pore-pressure perturbations, including ruptures growing beyond the perturbed area. Based on these concepts, they suggest that the largest earthquake magnitude scales with $\Delta V^{3/2}$ which they find consistent with the observed maximum magnitudes of injection-induced earthquakes over a broad range of injected volumes.

Although successful in a few cases, volume-based approaches fail to reproduce the post-injection seismicity that is typically observed after the stimulation experiments (see Section 6.2), when no further fluid volume is injected. Shapiro et al. [167] analyze the case histories of induced seismicity and find that the maximum seismic moment scales with the logarithm of the elapsed time from the beginning of the fluid injection to the earthquake occurrence, Δt. The scaling also applies to the case of Pohang EGS, and it includes post-injection seismicity since it describes the continuing growth of the stimulated rock volume after the termination of injection. However, Δt is unknown a priori; therefore, the authors suggest including the scaling in real-time monitoring of the seismogenic response and a probabilistic approach (see Section 6.5.3).

The approaches above neglect the role of rock hydraulic conductivity. In the case of a highly conductive rock formation of infinite extension, an arbitrary amount of

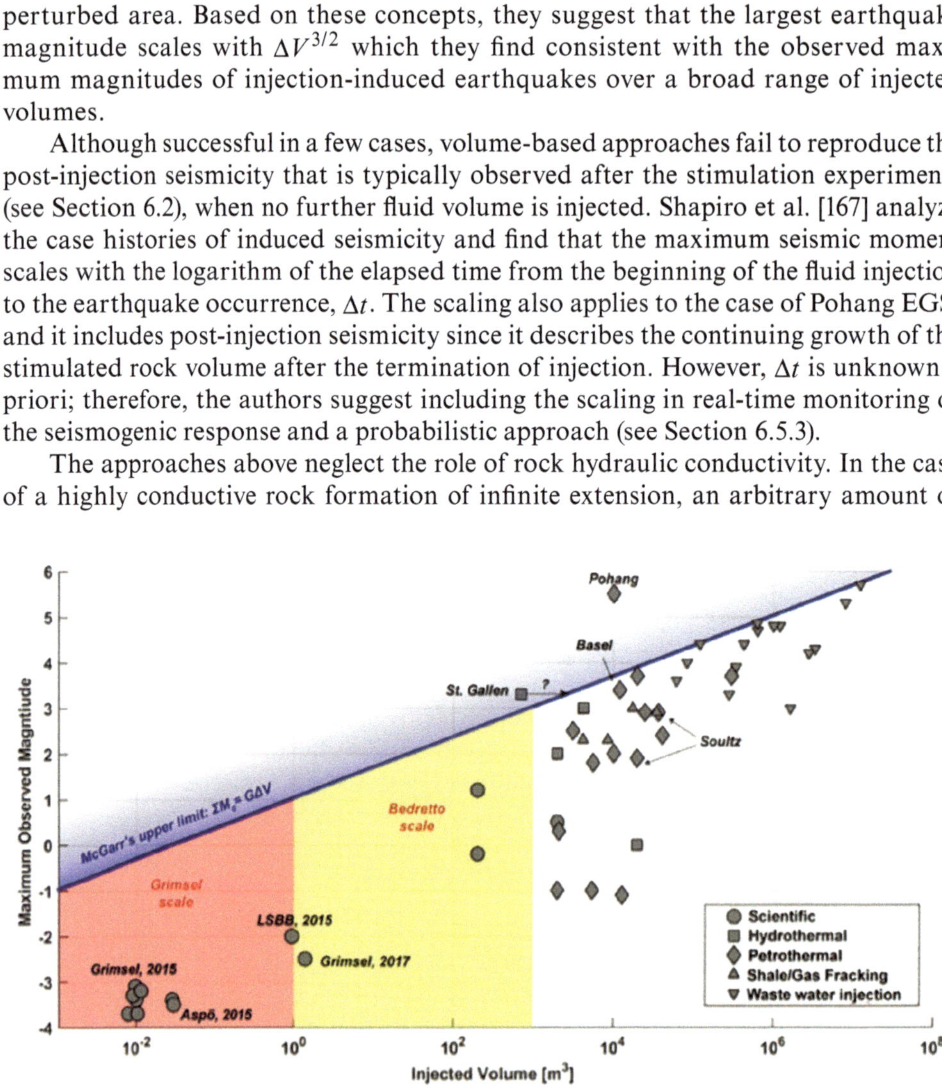

Figure 6.7 Largest earthquake magnitude versus injected volume for a diverse collection of fluid-injection projects. The upper limit on earthquake magnitude proposed by McGarr [75] is shown. Reprinted from Ref. [165], with permission from Elsevier.

fluid volume can be injected without significantly increasing the in-situ fluid pressure and thus without seismic energy release. In contrast with that, a number of authors proposed the hypothesis that the diffusive process of pore-pressure relaxation in porous (or fractured) saturated rocks is a dominant triggering mechanism of injection-induced microseismicity [168–172]. This approach includes the assumption that "the front of significant changes of the medium propagates behind the faster triggering front of earlier microseismic events" [170]. The reservoir hydraulic conductivity, either constant or pressure-dependent, defines the triggering front parabola that fits the space-time distribution of the seismic events [173].

Hummel and Muller [174] observed that the microseismic signature in the space/time diagram is not well fitted with the linear pressure diffusion equation, but they found a correlation with the maximum pressure gradient diffusion. The triggering front in space and time corresponds to the point where pressure perturbation exceeds zero for the first time, while the largest change in pore fluid pressure fits the correspondingly narrow band of increased event density. Linking seismicity with a sharp variation of pressure, namely a high pressure gradient, could explain the observation that, after shut-in and bleed-off, seismic activity took place mainly at the periphery of the stimulated volume [67].

Analytical estimates of pressure migration in a homogeneous domain with consequent direct effective stress reduction on predefined orientations have been adopted to model the observed seismicity at Basel during the co- and post-injection stages [175]. However, this simple model can only partly explain the observed seismicity.

Analytical models have the value of identifying the leading processes and parameters. However, the underlying physics is often too simplified because the maximum possible magnitude cannot be a function of the pressurized volume or the overpressure evolution alone. For this reason, the proposed scaling is often unsuccessful in predicting the observed seismicity, especially in the case of post-injection seismicity. In a critically stressed rock, induced seismicity largely depends on initial stress conditions, geometry, the presence of heterogeneity and the combination of effects.

6.5.2.2 Numerical approach

Numerical approaches attempt to predict induced seismicity by modeling the associated physical processes (Section 6.4) in the context of geomechanical modeling. Seismicity is typically not simulated as a dynamic process, but the occurrence of failure conditions is estimated from the evaluated stress. The latter requires the assumption of an initial stress field and a rock failure criterion, as described in Section 6.3.2. Numerous studies have modeled induced seismicity due to fluid injection, not only associated with EGS but also with other industrial activities, e.g., CO_2 sequestration, gas storage and hydraulic fracturing operations. The different studies mostly differ in the adopted geomechanical model and the simulated processes. When complex geometries and processes are considered, fully coupled numerical simulations have been employed, which have promoted the development of new methods, capabilities and computer software in the last 20 years (see also the reviews by Refs. [176] and [177]).

In the field of geomechanical models, a general distinction can be made based on the way the fractured reservoir is represented: some models use equivalent continuum media and others use discrete fracture elements. Continuum-based simulations consider a homogeneous porous reservoir without accounting for fault elements (e.g., [178]) or introduce one or more fault zones into the less permeable and stiffer rock matrix (e.g., [179]). The latter approach may be valid considering that a large number of faults are generally located in a narrow, highly damaged zone [180] and that, in some EGS projects, flow is observed to occur in fault zones with thicknesses of ten meters or more [60]. While continuum modeling is more appropriate for modeling volumetric poroelastic and thermoelastic effects, it often does not incorporate shear-slip activation and associated shear stress transfer. Some authors introduce the presence of a few fractures represented as explicit interface slipping elements into the rock matrix (e.g., [181]). Yoo et al. [182] and Kim et al. [183] performed HM modeling of the hydraulic stimulations at the Pohang EGS, where the shear slip and dilation are simulated by applying frictional plastic strain softening. Rutqvist et al. [184] model injection-induced slip on a pre-existing fault and estimate the magnitude of the associated seismic event according to Eq. 6.1 in the context of shale gas fracturing. However, the demanding computational requirements limit the number of fractures, physical processes, or geometry extension and space dimensions that can be numerically simulated.

Discrete fracture modeling explicitly considers the fluid flow and/or deformation at the fractures with a given fracture network geometry, often assuming an impermeable rock matrix. A set of pre-existing fractures is stochastically or deterministically generated, and fluid flow is numerically simulated with the hydro-mechanical interaction consisting of the opening or shearing of fractures. Taking into account the Coulomb failure criterion, shear failure is associated with the occurrence of seismicity, and a friction weakening mechanism is introduced to reproduce earthquake behavior (e.g., [181,185]). While the method allows for a faster representation of overpressure on the failure activation and of slip interactions due to slip stress transfer, it neglects the contribution of the host rock with respect to pressure and temperature diffusion and to volumetric poroelastic and thermoelastic effects.

The above-mentioned works, applied to reproduce either real or synthetic cases, have the value of contributing to a deeper understanding of the controlling processes and of the effects of reservoir characteristics. For example, thermo-mechanical effects (Section 6.4.2) have proven to be relevant in the long term [113,186,187]. Poroelastic effects are key, and they should not be neglected (see Section 6.4.1). They propagate ahead of the pressure front, and they may induce earthquakes in the post-injection stage, when pressure buildup drops [188–191]. Co-seismic static stress changes (Section 6.4.4) partially explain the observed induced seismicity [152,192]. De Simone et al. [114] show that only the combination of slip-induced stress redistribution and poro-thermo-elastic effects may explain the seismicity observed in the post-injection period in a synthetic example. Similar considerations have been proposed by Yeo [193], which proposes a combination of pore-pressure increase, earthquake interactions and fault weakening to explain the delayed seismicity registered at the Pohang geothermal field (Figure 6.8). The shear stress change may also be the result of aseismic slip [194,195], as measured in decameter-scale laboratory and field experiments [154,196,197]. In contrast to direct poromechanical and thermo-mechanical effects, which are likely to govern the reservoir response in the short term (hours to days) and intermediate periods (days to months), respectively, the chemical effects described in Section 6.4.3

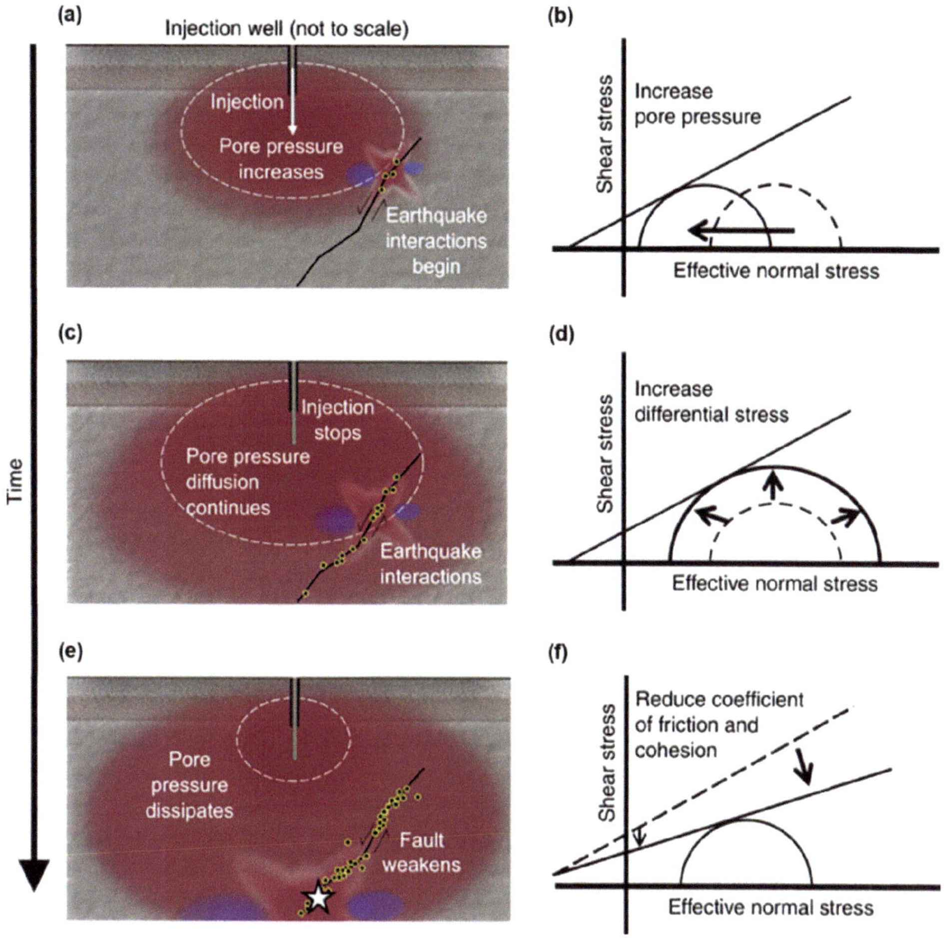

Figure 6.8 Conceptual illustrations of mechanisms triggering seismicity in Pohang EGS (South Korea). Reprinted from Ref. [193] under Creative Common License http://creativecommons.org/licenses/by/4.0/.

can become dominant only after months to years of fluid injection [104,126,131]. Nevertheless, the nature, rates and extents of geochemical processes are poorly constrained and broadly known to be case-specific, complicating their assessment [160].

A number of studies have highlighted the impact of reservoir characteristics, such as hydraulic, mechanical and frictional properties, the orientation of faults with respect to the stress regime, fracture distribution and connection, fault location with respect to the pressurized region, and fault offset (e.g., [117,198,199]), as well as the impact of injection rate and shut-in procedure on induced seismicity (e.g., [200,201]). The numerical approach has the potential to perform as a forecasting model because the physical processes are accurately described. Moreover, physics-based models offer the possibility of identifying the leading processes and exploring the sensitivity of seismic response to site characteristics and operation strategies. Numerical modeling offers the possibility of considering the combination and interaction of several mechanisms. The main drawback resides in the difficulties in properly reproducing the processes and in characterizing the numerous parameters and in-situ conditions.

6.5.3 Hybrid models

The hybrid approach consists of a combination of physics-based models with statistical models, taking the advantages of both approaches. Some authors attempted to estimate the expected magnitudes of induced earthquakes, assuming that pore-pressure diffusion is the governing mechanism leading to the triggering of seismic events and statistically assigning critical values of pore pressure to a random set of pre-existing fractures (e.g., [202]). Parotidis et al. [203] assumed that the pore-pressure values leading to earthquakes approximately follow a rectangular probability distribution. Shapiro et al. [204] find that the process of pressure diffusion in a poroelastic medium with randomly distributed sub-critical cracks obeying a G-R relation well explains the observations of the temporal magnitude distribution of fluid-induced seismicity. Langenbruch and Shapiro [205] proposed a modified Omori law for fluid-induced seismicity to estimate the decay rate in dependence on parameters of injection, reservoir rock and the strength of pre-existing fractures in a reservoir.

Based on the approach proposed by Rothert and Shapiro [202], Bachmann et al. [206] analyze the spatio-temporal variability of the b-value in the G-R relation and compare the observed magnitude distribution with the results of a minimalistic model of pore-pressure evolution that relates magnitudes to differential stress. They observe that the decrease of b-values with increasing distance from the injection point might be caused by a decrease in pore-pressure. This leads to an increase in the probability of a large magnitude event with distance and time. Similarly, Goertz-Allmann and Wiemer [207] use pressure diffusion in a homogeneous medium to estimate seismicity triggering based on the Coulomb failure criterion. The earthquake magnitude is modeled with a semiprobabilistic approach based on drawing a random magnitude from a Gutenberg–Richter distribution with b-value proportional to the differential stress. The model is successfully applied to forecast the observed seismicity at the Basel EGS. Successively, the approach is further developed to incorporate more complex and realistic processes, such as non-linear pressure diffusion and permeability enhancement during stimulation [208] and hydro-mechanical pressure and stress changes, with slip-induced stress redistribution estimated via numerical simulations [209].

The hybrid approach appears promising in the light of building a real-time probabilistic seismic hazard analysis (PSHA) during injection. In fact, fully coupled simulations are too computationally expensive, provided that they also need to be calibrated according to the observations. On the contrary, a hybrid approach, combining simplified physical/mathematical models with a probabilistic seismic analysis, has the potential to provide a quick assessment of seismic hazards.

6.6 CLOSURE

Induced seismicity represents a paramount challenge for scientific researchers and operators because the associated loss of confidence and safety risk jeopardize the development of geo-energy technologies, like EGS. The mitigation of large seismic events can only be achieved when the underlying processes are well understood and accurately predicted. In this light, numerical simulations are fundamental to modeling the interactions between fluid flow, stress variation, heat transport and geochemical reactions. The main difficulty in adopting this approach lies in the accurate reproduction of the physical processes and the proper characterization of the reservoir. Recent studies have made significant contributions to understanding the specific and combined effects of these processes [114,188,193,194,196]. Additional efforts should, however, be made in several directions. There is a need for new approaches and computing advances to simulate the coupled THMC processes under different scenarios that account for the geological uncertainties of the reservoir structure and of the mechanical, thermal, hydraulic and chemical parameters [160]. The scaling of the temporal delay of the events with the formation parameters is still unclear, and it requires further investigations. Efforts to improve the in-situ characterization, as well as laboratory and field measurements to capture the triggering processes, are valuable. The recent in-situ decameter-scale or hundred-meter-scale experiments with advanced monitoring systems (e.g., the Swiss underground laboratory of Bedretto and Grimsel [165], see Amann et al. [210] for a comprehensive review) are providing insightful contributions in this sense. Little progress has been made so far on understanding the geochemical control of induced seismicity, which can be attributed to their complexity and prolonged impacts. Efforts should continue to fill in this knowledge gap on a long-term and large-scale basis through coupled THMC modeling integrated with a variety of well-established characterization and monitoring approaches.

REFERENCES

[1] P. Talwani, L. Chen and K. Gahalaut, "Seismogenic permeability, ks," *Journal of Geophysical Research,* vol. 112, B07309, 2007.

[2] H. K. Gupta, "A review of recent studies of triggered earthquakes by artificial water reservoirs with special emphasis on earthquakes in Koyna, India," *Earth-Science Reviews,* vol. 58, p. 279–310, 2002.

[3] A. J. Mendecki, Ed., *Seismic Monitoring in Mines*, Springer Dordrecht, 1997.

[4] A. McGarr, "Seismic moments and volume changes," *Journal of Geophysical Research,* vol. 81, p. 1487–1494, 1976.

[5] P. Segall, "Earthquakes triggered by fluid extraction," *Geology,* vol. 17, p. 942–946, 1989.

[6] J. R. Grasso, "Mechanics of seismic instabilities induced by the recovery of hydrocarbons," *Pure and Applied Geophysics PAGEOPH,* vol. 139, p. 507–534, 1992.

[7] J. Suckale, "Induced seismicity in hydrocarbon fields," in *Advances in Geophysics,* Vol. 51, Elsevier, 2009, p. 55–106.

[8] C. B. Raleigh, J. H. Healy and J. D. Bredehoeft, "Faulting and crustal stress at Rangely, Colorado," in *Flow and Fracture of Rocks*, Geophysical Monograph Series,vol. 16, edited by H. C. Heard et al., AGU, Washington, DC., 1972, p. 275–284.

[9] J. H. Healy, W. W. Rubey, D. T. Griggs and C. B. Raleigh, "The Denver earthquakes," *Science,* vol. 161, p. 1301–1310, 1968.

[10] K. F. Evans, A. Zappone, T. Kraft, N. Deichmann and F. Moia, "A survey of the induced seismic responses to fluid injection in geothermal and CO_2 reservoirs in Europe," *Geothermics,* vol. 41, p. 30–54, 2012.

[11] A. McGarr, D. Simpson, L. Seeber and W. Lee, "Case histories of induced and triggered seismicity," *International Geophysics Series,* vol. 81, p. 647–664, 2002.

[12] T. Dahm, S. Cesca, S. Hainzl, T. Braun and F. Krueger, "Discrimination between induced, triggered, and natural earthquakes close to hydrocarbon reservoirs: A probabilistic approach based on the modeling of depletion-induced stress changes and seismological source parameters," *Journal of Geophysical Research: Solid Earth,* vol. 120, p. 2491–2509, 2015.
[13] L. Buijze, L. van Bijsterveldt, H. Cremer, B. Paap, H. Veldkamp, B. B. T. Wassing, J.-D. van Wees, G. C. N. van Yperen, J. H. ter Heege and B. Jaarsma, "Review of induced seismicity in geothermal systems worldwide and implications for geothermal systems in the Netherlands," *Netherlands Journal of Geosciences,* vol. 98, e13, 2019.
[14] W. L. Ellsworth, "Injection-induced earthquakes," *Science,* vol. 341, 1225942, 2013.
[15] F. Grigoli, S. Cesca, E. Priolo, A. P. Rinaldi, J. F. Clinton, T. A. Stabile, B. Dost, M. G. Fernandez, S. Wiemer and T. Dahm, "Current challenges in monitoring, discrimination, and management of induced seismicity related to underground industrial activities: A European perspective," *Reviews of Geophysics,* vol. 55, p. 310–340, 2017.
[16] M. W. Hitzman, D. D. Clarke, E. Detournay, J. H. Dieterich, D. K. Dillon, S. J. Greenet al., *Induced Seismicity Potential in Energy Technologies,* National Academies Press, Washington, DC, 2012.
[17] E. L. Majer, R. Baria, M. Stark, S. Oates, J. Bommer, B. Smith and H. Asanuma, "Induced seismicity associated with Enhanced Geothermal Systems," *Geothermics,* vol. 36, p. 185–222, 2007.
[18] A. Zang, V. Oye, P. Jousset, N. Deichmann, R. Gritto, A. McGarr, E. Majer and D. Bruhn, "Analysis of induced seismicity in geothermal reservoirs – An overview," *Geothermics,* vol. 52, p. 6–21, 2014.
[19] G. R. Foulger, M. P. Wilson, J. G. Gluyas, B. R. Julian and R. J. Davies, "Global review of human-induced earthquakes," *Earth-Science Reviews,* vol. 178, p. 438–514, 2018.
[20] F. Batini, R. Console and G. Luongo, "Seismological study of Larderello — Travale geothermal area," *Geothermics,* vol. 14, p. 255–272, 1985.
[21] F. Moia, "Individuazione ed applicazione di metodologie di monitoraggio di possibili fughe di CO_2 dai serbatoi di stoccaggio," Report No. 08001015. Cesi Ricerca (Milan), 2008.
[22] F. Batini, G. M. Cameli, E. Carabelli and A. Fiordelisi, "Seismic monitoring in Italian geothermal areas II: Seismic activity in the geothermal fields during exploitation," in *Proceedings of Second DOE-ENEL Workshop on Cooperative Research in Geothermal Energy,* Report LBL-11555, Lawrence Berkeley Laboratory, Berkeley, CA, October 1980.
[23] A. Mazzoldi, A. Borgia, M. Ripepe, E. Marchetti, G. Ulivieri, M. della Schiava and C. Allocca, "Faults strengthening and seismicity induced by geothermal exploitation on a spreading volcano, Mt. Amiata, Italia," *Journal of Volcanology and Geothermal Research,* vol. 301, p. 159–168, 2015.
[24] T. Braun, S. Cesca, D. Kühn, A. Martirosian-Janssen and T. Dahm, "Anthropogenic seismicity in Italy and its relation to tectonics: State of the art and perspectives," *Anthropocene,* vol. 21, p. 80–94, 2018.
[25] D. H. Oppenheimer, "Extensional tectonics at The Geysers Geothermal Area, California," *Journal of Geophysical Research,* vol. 91, p. 11463, 1986.
[26] D. Eberhart-Phillips and D. H. Oppenheimer, "Induced seismicity in The Geysers Geothermal Area, California," *Journal of Geophysical Research: Solid Earth,* vol. 89, p. 1191–1207, 1984.
[27] A. Mossop and P. Segall, "Subsidence at The Geysers Geothermal Field, N. California from a comparison of GPS and leveling surveys," *Geophysical Research Letters,* vol. 24, p. 1839–1842, 1997.
[28] E. L. Majer and J. E. Peterson, "The impact of injection on seismicity at The Geysers, California Geothermal Field," *International Journal of Rock Mechanics and Mining Sciences,* vol. 44, p. 1079–1090, 2007.

[29] M. A. Stark, "Microearthquakes–a tool to track injected water in The Geysers reservoir," *Monograph on the Geyser Geothermal Field, Special Report*, vol. 17, p. 111–117, 1992.
[30] E. L. Majer, K. Freeman, L. Johnson, S. Jarpe, K. Nihei, C. Hartline, M. Walters and M. Dellinger, *Monitoring the Effect of Injection of Fluids from the Lake County Pipeline on Seismicity at The Geysers, California, Geothermal Field*, Final Report. Lawrence Berkeley National Laboratory, Calpine Corporation, Lake County, CA, 2017.
[31] B. Smith, J. Beall and M. Stark, "Induced seismicity in the SE Geysers field, California, USA," in *Proceedings, World Geothermal Congress 2000*, Kyushu - Tohoku, Japan, May 28–June 10, 2000.
[32] M. A. Stark, W. T. Box Jr, J. J. Beall, K. P. Goyal and A. S. Pingol, "The Santa Rosa–geysers recharge project, geysers geothermal field, California, USA," *Proceedings, World Geothermal Congress*, Antalya, 24–29 April 2005, p. 24–29, 2005.
[33] P. Martínez-Garzón, G. Kwiatek, H. Sone, M. Bohnhoff, G. Dresen and C. Hartline, "Spatiotemporal changes, faulting regimes, and source parameters of induced seismicity: A case study from The Geysers geothermal field," *Journal of Geophysical Research: Solid Earth*, vol. 119, p. 8378–8396, 2014.
[34] K. Aki, M. Fehler, R. L. Aamodt, J. N. Albright, R. M. Potter, C. M. Pearson and J. W. Tester, "Interpretation of seismic data from hydraulic fracturing experiments at the Fenton Hill, New Mexico, hot dry rock geothermal site," *Journal of Geophysical Research: Solid Earth*, vol. 87, p. 936–944, 1982.
[35] W. S. Phillips, L. S. House and M. C. Fehler, "Detailed joint structure in a geothermal reservoir from studies of induced microearthquake clusters," *Journal of Geophysical Research: Solid Earth*, vol. 102, p. 11745–11763, 1997.
[36] R. J. Pine and A. S. Batchelor, "Downward migration of shearing in jointed rock during hydraulic injections," *International Journal of Rock Mechanics and Mining Sciences & Geomechanics Abstracts*, vol. 21, p. 249–263, 1984.
[37] H. G. Richards, R. H. Parker, A. S. P. Green, R. H. Jones, J. D. M. Nicholls, D. A. C. Nicol, M. M. Randall, S. Richards, R. C. Stewart and J. Willis-Richards, "The performance and characteristics of the experimental hot dry rock geothermal reservoir at Rosemanowes, Cornwall (1985–1988)," *Geothermics*, vol. 23, p. 73–109, 1994.
[38] R. Parker, "The Rosemanowes HDR project 1983–1991," *Geothermics*, vol. 28, p. 603–615, 1999.
[39] S. Sasaki, "Characteristics of microseismic events induced during hydraulic fracturing experiments at the Hijiori hot dry rock geothermal energy site, Yamagata, Japan," *Tectonophysics*, vol. 289, p. 171–188, 1998.
[40] H. Kaieda, "Multiple reservoir creation and evaluation in the Ogachi and Hijiori HDR projects, Japan," in *Proceedings* of the *World Geothermal Congress*, Melbourne, 2015.
[41] A. J. Jupe, A. S. P. Green and T. Wallroth, "Induced microseismicity and reservoir growth at the Fjällbacka hot dry rocks project, Sweden," *International Journal of Rock Mechanics and Mining Sciences & Geomechanics Abstracts*, vol. 29, p. 343–354, 1992.
[42] T. Wallroth, A. J. Jupe and R. H. Jones, "Characterisation of a fractured reservoir using microearthquakes induced by hydraulic injections," *Marine and Petroleum Geology*, vol. 13, p. 447–455, 1996.
[43] H. Kaieda, S. Sasaki and D. Wyborn, "Comparison of characteristics of micro-earthquakes observed during hydraulic stimulation operations in Ogachi, Hijiori and Cooper Basin HDR projects," in *Proceedings of the World Geothermal Congress*, Bali, 25–29 April 2010.
[44] J. J. McGuire, R. B. Lohman, R. D. Catchings, M. J. Rymer and M. R. Goldman, "Relationships among seismic velocity, metamorphism, and seismic and aseismic fault slip in the Salton Sea Geothermal Field region," *Journal of Geophysical Research: Solid Earth*, vol. 120, p. 2600–2615, 2015.

[45] J. Groos, J. Zeiß, M. Grund and J. Ritter, "Microseismicity at two geothermal power plants at Landau and Insheim in the Upper Rhine Graben, Germany," *Geophysical Research Abstracts* vol. 15. , 2013.

[46] L. Küperkoch, K. Olbert and T. Meier, "Long-term monitoring of induced seismicity at the Insheim Geothermal Site, Germany," *Bulletin of the Seismological Society of America,* vol. 108, no. 6, p. 3668–3683, 2018.

[47] J. Albaric, V. Oye, N. Langet, M. Hasting, I. Lecomte, K. Iranpour, M. Messeiller and P. Reid, "Monitoring of induced seismicity during the first geothermal reservoir stimulation at Paralana, Australia," *Geothermics,* vol. 52, p. 120–131, 2014.

[48] G. Bjornsson and I. S. O. R. Iceland GeoSurvey, "Reservoir conditions at 3–6 km depth in the Hellisheidi geothermal field, SW-Iceland, estimated by deep drilling, cold water injection and seismic monitoring," in *Proceedings, Twenty-Ninth Workshop on Geothermal Reservoir Engineering Stanford University*, Stanford, CA, January 26–28, 2004.

[49] Ó. G. Flóvenz, K. Ágústsson, E. Á. Guðnason and S. Kristjánsdóttir, "Reinjection and induced seismicity in geothermal fields in Iceland," in *Proceedings World Geothermal Congress*, Melbourne, 19–25 April 2015.

[50] J. J. Bommer, S. Oates, J. M. Cepeda, C. Lindholm, J. Bird, R. Torres, G. Marroquín and J. Rivas, "Control of hazard due to seismicity induced by a hot fractured rock geothermal project," *Engineering Geology,* vol. 83, p. 287–306, 2006.

[51] G. Kwiatek, F. Bulut, M. Bohnhoff and G. Dresen, "High-resolution analysis of seismicity induced at Berlín geothermal field, El Salvador," *Geothermics,* vol. 52, p. 98–111, 2014.

[52] H. Asanuma, N. Soma, H. Kaieda, Y. Kumano, T. Izumi, K. Tezuka, H. Niitsuma and D. Wyborn, "Microseismic monitoring of hydraulic stimulation at the Australian HDR project in Cooper Basin," in *Proceedings World Geothermal Congress*, Antalya, 24–29 April 2005.

[53] S. Baisch, R. Weidler, R. Voros, D. Wyborn and L. de Graaf, "Induced seismicity during the stimulation of a geothermal HFR reservoir in the Cooper Basin, Australia," *Bulletin of the Seismological Society of America,* vol. 96, p. 2242–2256, 2006.

[54] S. Baisch, R. Voros, R. Weidler and D. Wyborn, "Investigation of fault mechanisms during geothermal reservoir stimulation experiments in the Cooper Basin, Australia," *Bulletin of the Seismological Society of America,* vol. 99, p. 148–158, 2009.

[55] S. Baisch, E. Rothert, H. Stang, R. Vörös, C. Koch and A. McMahon, "Continued geothermal reservoir stimulation experiments in the Cooper Basin (Australia)," *Bulletin of the Seismological Society of America,* vol. 105, p. 198–209, 2015.

[56] H.-G. Holl and C. Barton, "Habanero field—structure and state of stress," in *Proceedings World Geothermal Congress*, Melbourne, 19–25 April 2015.

[57] R. Baria, S. Michelet, J. Baumgärtner, B. Dyer, J. Nicholls, T. Hettkamp, D. Teza, N. Soma, H. Asanuma, J. Garnish andet al., "Creation and mapping of 5000 m deep HDR/HFR reservoir to produce electricity," *Proceedings* World Geothermal Congress 2005, Antalya, 24–29 April, 2005.

[58] A. Genter, D. Fritsch, N. Cuenot, J. Baumgartner and J. Graff, "Overview of the current activities of the European EGS Soultz project: from exploration to electricity production," in *Proceedings, Thirty-Fourth Workshop on Geothermal Reservoir Engineering*, Stanford University, Stanford, CA, February 9–11, 2009.

[59] K. F. Evans, A. Genter and J. Sausse, "Permeability creation and damage due to massive fluid injections into granite at 3.5 km at Soultz: 1. Borehole observations," *Journal of Geophysical Research: Solid Earth,* vol. 110, B04203, 2005.

[60] A. Genter, K. Evans, N. Cuenot, D. Fritsch and B. Sanjuan, "Contribution of the exploration of deep crystalline fractured reservoir of Soultz to the knowledge of enhanced geothermal systems (EGS)," *Comptes Rendus Geoscience,* vol. 342, p. 502–516, 2010.

[61] F. H. Cornet, T. Bérard and S. Bourouis, "How close to failure is a granite rock mass at a 5 km depth?", *International Journal of Rock Mechanics and Mining Sciences,* vol. 44, p. 47–66, 2007.
[62] R. Baria, J. Baumgaerner, A. Gérard and R. Jung, "European hot dry rock geothermal research programme 1996–1997," *Final Report to the European Commission, Brussels, Belgium*, vol. 192, 1997.
[63] R. Baria, S. Michelet, J. Baumgaertner, B. Dyer, A. Gerard, J. Nicholls, T. Hettkamp, D. Teza, N. Soma, H. Asanuma andet al., "Microseismic monitoring of the world's largest potential HDR reservoir," in *Proceedings, Twenty-Ninth Workshop on Geothermal Reservoir Engineering*, Stanford University, Stanford, CA, January 26–28, 2004.
[64] L. Dorbath, N. Cuenot, A. Genter and M. Frogneux, "Seismic response of the fractured and faulted granite of Soultz-sous-Forêts (France) to 5 km deep massive water injections," *Geophysical Journal International,* vol. 177, p. 653–675, 2009.
[65] M. O. Häring, U. Schanz, F. Ladner and B. C. Dyer, "Characterisation of the Basel 1 enhanced geothermal system,"*Geothermics,* vol. 37, p. 469–495, 2008.
[66] N. Deichmann and J. Ernst, "Earthquake focal mechanisms of the induced seismicity in 2006 and 2007 below Basel (Switzerland)," *Swiss Journal of Geosciences,* vol. 102, 457, 2009.
[67] N. Deichmann and D. Giardini, "Earthquakes Induced by the Stimulation of an Enhanced Geothermal System below Basel (Switzerland)," *Seismological Research Letters,* vol. 80, p. 784–798, 2009.
[68] J. Ripperger, P. Kästli, D. Fäh and D. Giardini, "Ground motion and macroseismic intensities of a seismic event related to geothermal reservoir stimulation below the city of Basel-observations and modelling," *Geophysical Journal International,* vol. 179, p. 1757–1771, 2009.
[69] F. Grigoli, S. Cesca, A. P. Rinaldi, A. Manconi, J. A. López-Comino, J. F. Clinton, R. Westaway, C. Cauzzi, T. Dahm and S. Wiemer, "The November 2017 M w 5.5 Pohang earthquake: A possible case of induced seismicity in South Korea," *Science,* vol. 360, p. 1003–1006, 2018.
[70] K.-H. Kim, J.-H. Ree, Y. Kim, S. Kim, S. Y. Kang and W. Seo, "Assessing whether the 2017 M w 5.4 Pohang earthquake in South Korea was an induced event," *Science,* vol. 360, p. 1007–1009, 2018.
[71] S. Ge, D. Giardini, W. Ellsworth, T. Shimamoto and J. Townend, *Overseas Research Advisory Committee Report on the Pohang Earthquake*, The Geological Society of Korea, Seoul, 2019.
[72] S. Park, K.-I. Kim, L. Xie, H. Yoo, K.-B. Min, M. Kim, B. Yoon, K. Y. Kim, G. Zimmermann, F. Guinot and P. Meier, "Observations and analyses of the first two hydraulic stimulations in the Pohang geothermal development site, South Korea," *Geothermics,* vol. 88, p. 101905, 2020.
[73] W. L. Ellsworth, D. Giardini, J. Townend, S. Ge and T. Shimamoto, "Triggering of the Pohang, Korea, earthquake (M w 5.5) by enhanced geothermal system stimulation," *Seismological Research Letters,* vol. 90, p. 1844–1858, 2019.
[74] Korean Government Commission, *Summary Report of the Korean Government Commission on Relations Between the 2017 Pohang Earthquake and EGS Project*, Geological Society of Korea, Seoul, 2019.
[75] A. McGarr, "Maximum magnitude earthquakes induced by fluid injection," *Journal of Geophysical Research: Solid Earth,* vol. 119, p. 1008–1019, 2014.
[76] H. Kanamori and E. E. Brodsky, "The physics of earthquakes," *Reports on Progress in Physics,* vol. 67, p. 1429, 2004.
[77] J. Brune and W. Thatcher, Strength and Energetics of Active Fault Zones. In *International Handbook of Earthquake and Engineering Seismology*, Part A, *Volume 81A* Eds. Lee et al. Elsevier,p. 569–588, 2002.

[78] J. A. Steketee, "Some geophysical applications of the elasticity theory of dislocations," *Canadian Journal of Physics,* vol. 36, p. 1168–1198, 1958.
[79] C. F. Richter, "An instrumental earthquake magnitude scale*," *Bulletin of the Seismological Society of America,* vol. 25, p. 1–32, 1935.
[80] H. Kanamori, "The energy release in great earthquakes," *Journal of Geophysical Research,* vol. 82, p. 2981–2987, 1977.
[81] H. Kanamori, "Quantification of earthquakes," *Nature,* vol. 271, p. 411–414, 1978.
[82] T. C. Hanks and H. Kanamori, "A moment magnitude scale," *Journal of Geophysical Research,* vol. 84, p. 2348–2350, 1979.
[83] R. F. Mereu, "A study of the relations between ML, Me, Mw, apparent stress, and fault aspect ratio," *Physics of the Earth and Planetary Interiors,* vol. 298, p. 106278, 2020.
[84] D. L. Wells and K. J. Coppersmith, "New empirical relationships among magnitude, rupture length, rupture width, rupture area, and surface displacement," *Bulletin of the Seismological Society of America,* vol. 84, p. 974–1002, 1994.
[85] J. R. Rice and M. P. Cleary, "Some basic stress diffusion solutions for fluid-saturated elastic porous media with compressible constituents," *Reviews of Geophysics,* vol. 14, p. 227, 1976.
[86] R. H. Sibson, "Interactions between temperature and pore-fluid pressure during earthquake faulting and a mechanism for partial or total stress relief," *Nature Physical Science,* vol. 243, p. 66–68, 1973.
[87] T. Utsu, "Statistical features of seismicity," *International Geophysics Series,* vol. 81, p. 719–732, 2002.
[88] F. Omori, "On the after-shocks of earthquakes" *Journal of the College of Science,* Imperial University of Tokyo Japan, vol. 7, part 2, p. 111–200, 1894.
[89] T. Utsu, "A statistical study on the occurrence of aftershocks," *The Geophysical Magazine,* vol. 30, p. 521–605, 1961.
[90] B. Gutenberg and C. F. Richter, "Earthquake magnitude, intensity, energy, and acceleration*," *Bulletin of the Seismological Society of America,* vol. 32, p. 163–191, 1942.
[91] C. E. Bachmann, S. Wiemer, J. Woessner and S. Hainzl, "Statistical analysis of the induced Basel 2006 earthquake sequence: introducing a probability-based monitoring approach for Enhanced Geothermal Systems," *Geophysical Journal International,* vol. 186, p. 793–807, 2011.
[92] Y. Ogata, "Statistical models for earthquake occurrences and residual analysis for point processes," *Journal of the American Statistical Association,* vol. 83, p. 9–27, 1988.
[93] P. A. Reasenberg and L. M. Jones, "Earthquake hazard after a mainshock in California," *Science,* vol. 243, p. 1173–1176, 1989.
[94] J. Woessner, A. Christophersen, J. D. Zechar and D. Monelli, "Building self-consistent, short-term earthquake probability (STEP) models: Improved strategies and calibration procedures," *Annals of Geophysics,* vol. 53(3), pp. 141–154, 2010.
[95] E. M. Anderson, "The dynamics of faulting," *Transactions of the Edinburgh Geological Society,* vol. 8, p. 387–402, 1905.
[96] O. Heidbach, M. Rajabi, X. Cui, K. Fuchs, B. Müller, J. Reinecker, K. Reiter, M. Tingay, F. Wenzel, F. Xie, M. O. Ziegler, M.-L. Zoback and M. Zoback, "The World Stress Map database release 2016: Crustal stress pattern across scales," *Tectonophysics,* vol. 744, p. 484–498, 2018.
[97] M. D. Zoback, C. A. Barton, M. Brudy, D. A. Castillo, T. Finkbeiner, B. R. Grollimund, D. B. Moos, P. Peska, C. D. Ward and D. J. Wiprut, "Determination of stress orientation and magnitude in deep wells," *International Journal of Rock Mechanics and Mining Sciences,* vol. 40, p. 1049–1076, 2003.
[98] J. C. Jaeger, N. G. W. Cook and R. Zimmerman, *Fundamentals of Rock Mechanics,* John Wiley & Sons, 2009.

[99] J. B. Clark, "A hydraulic process for increasing the productivity of wells," *Journal of Petroleum Technology,* vol. 1, p. 1–8, 1949.
[100] J. Byerlee, *Friction of Rocks*, Vol. 116, Birkhäuser, Basel, 1978, p. 615–626.
[101] F. H. Cornet, J. Helm, H. Poitrenaud and A. Etchecopar, "Seismic and aseismic slips induced by large-scale fluid injections," in *Seismicity Associated with Mines, Reservoirs and Fluid Injections*, Ed. Talebi S., Pageoph Topical Volumes Birkhäuser, Basel, 1997, p. 563–583.
[102] H. Fossen, *Structural Geology*, Cambridge University Press, New York, 2016.
[103] D. Turcotte and G. Schubert, *Geodynamics*, Cambridge University Press, New York, 2014.
[104] A. Ghassemi, "A review of some rock mechanics issues in geothermal reservoir development," *Geotechnical and Geological Engineering,* vol. 30, p. 647–664, 2012.
[105] P. Segall and S. D. Fitzgerald, "A note on induced stress changes in hydrocarbon and geothermal reservoirs," *Tectonophysics,* vol. 289, p. 117–128, 1998.
[106] C. A. J. Wibberley and T. Shimamoto, "Earthquake slip weakening and asperities explained by thermal pressurization," *Nature,* vol. 436, p. 689–692, 2005.
[107] G. De Marsily, *Quantitative Hydrogeology*, Paris School of Mines, Fontainebleau, 1986.
[108] M. A. Biot, "General theory of three-dimensional consolidation," *Journal of Applied Physics,* vol. 12, p. 155–164, 1941.
[109] J. W. Rudnicki, "Eshelby's technique for analyzing inhomogeneities in geomechanics," in *Mechanics of Crustal Rocks*, Eds. Leroy, Y.M., Lehner, F.K., CISM Courses and Lectures, vol. 533, Springer Vienna, p. 43–72, 2011.
[110] H. Soltanzadeh and C. D. Hawkes, "Assessing fault reactivation tendency within and surrounding porous reservoirs during fluid production or injection," *International Journal of Rock Mechanics and Mining Sciences,* vol. 46, p. 1–7, 2009.
[111] H. Soltanzadeh and C. D. Hawkes, "Semi-analytical models for stress change and fault reactivation induced by reservoir production and injection," *Journal of Petroleum Science and Engineering,* vol. 60, p. 71–85, 2008.
[112] S. De Simone and J. Carrera, "Analytical solutions to coupled HM problems to highlight the nonlocal nature of aquifer storage," *Water Resources Research,* vol. 53, p. 9580–9599, 2017.
[113] S. De Simone, V. Vilarrasa, J. Carrera, A. Alcolea and P. Meier, "Thermal coupling may control mechanical stability of geothermal reservoirs during cold water injection," *Physics and Chemistry of the Earth, Parts A/B/C,* vol. 64, p. 117–126, 2013.
[114] S. De Simone, J. Carrera and V. Vilarrasa, "Superposition approach to understand triggering mechanisms of post-injection induced seismicity," *Geothermics,* vol. 70, p. 85–97, 2017.
[115] A. Zareidarmiyan, H. Salarirad, V. Vilarrasa, S. De Simone and S. Olivella, "Geomechanical response of fractured reservoirs," *Fluids,* vol. 3, p. 70, 2018.
[116] H. Wu, V. Vilarrasa, S. De Simone, M. Saaltink and F. Parisio, "Analytical solution to assess the induced seismicity potential of faults in pressurized and depleted reservoirs," *Journal of Geophysical Research: Solid Earth,* vol. 126, e2020JB020436, 2021.
[117] V. Vilarrasa, "The role of the stress regime on microseismicity induced by overpressure and cooling in geologic carbon storage," *Geofluids,* vol. 16, p. 941–953, 2016.
[118] J. Rutqvist and O. Stephansson, "The role of hydromechanical coupling in fractured rock engineering," *Hydrogeology Journal,* vol. 11, p. 7–40, 2003.
[119] K. F. Evans, H. Moriya, H. Niitsuma, R. H. Jones, W. S. Phillips, A. Genter, J. Sausse, R. Jung and R. Baria, "Microseismicity and permeability enhancement of hydrogeologic structures during massive fluid injections into granite at 3 km depth at the Soultz HDR site," *Geophysical Journal International,* vol. 160, p. 389–412, 2004.
[120] J. W. Rudnicki, "Alteration of regional stress by reservoirs and other inhomogeneities: stabilizing or destabilizing?" in *9th ISRM Congress*, Paris, 1999.

[121] M. A. Biot, "Thermoelasticity and irreversible thermodynamics," *Journal of Applied Physics,* vol. 27, p. 240–253, 1956.

[122] V. V. Palciauskas and P. A. Domenico, "Characterization of drained and undrained response of thermally loaded repository rocks," *Water Resources Research,* vol. 18, p. 281–290, 1982.

[123] R. W. Zimmerman, "Coupling in poroelasticity and thermoelasticity," *International Journal of Rock Mechanics and Mining Sciences,* vol. 37, p. 79–87, 2000.

[124] S. De Simone, J. Carrera and B. M. Gómez-Castro, "A practical solution to the mechanical perturbations induced by non-isothermal injection into a permeable medium," *International Journal of Rock Mechanics and Mining Sciences,* vol. 91, p. 7–17, 2017.

[125] I. R. Kivi, E. Pujades, J. Rutqvist and V. Vilarrasa, "Cooling-induced reactivation of distant faults during long-term geothermal energy production in hot sedimentary aquifers," *Scientific Reports,* vol. 12, p. 2065, 2022.

[126] S. N. Pandey, V. Vishal and A. Chaudhuri, "Geothermal reservoir modeling in a coupled thermo-hydro-mechanical-chemical approach: A review," *Earth-Science Reviews,* vol. 185, p. 1157–1169, 2018.

[127] B. J. Andre and H. Rajaram, "Dissolution of limestone fractures by cooling waters: Early development of hypogene karst systems," *Water Resources Research,* vol. 41, W01015, 2005.

[128] S. N. Pandey, A. Chaudhuri, H. Rajaram and S. Kelkar, "Fracture transmissivity evolution due to silica dissolution/precipitation during geothermal heat extraction," *Geothermics,* vol. 57, p. 111–126, 2015.

[129] D. L. Gallup and P. von Hirtz, "Control of silica-based scales in cooling and geothermal systems," in *Mineral Scales and Deposits*, Eds. Amjad Z. and Demadis K. D., Elsevier, Netherlands, pp. 573–582, 2015.

[130] A. C. Lasaga, J. M. Soler, J. Ganor, T. E. Burch and K. L. Nagy, "Chemical weathering rate laws and global geochemical cycles," *Geochimica et Cosmochimica Acta,* vol. 58, p. 2361–2386, 1994.

[131] J. Taron and D. Elsworth, "Thermal–hydrologic–mechanical–chemical processes in the evolution of engineered geothermal reservoirs," *International Journal of Rock Mechanics and Mining Sciences,* vol. 46, p. 855–864, 2009.

[132] Z. Jing, K. Watanabe, J. Willis-Richards and T. Hashida, "A 3-D water/rock chemical interaction model for prediction of HDR/HWR geothermal reservoir performance," *Geothermics,* vol. 31, p. 1–28, 2002.

[133] K. M. Kovac, T. Xu, K. Pruess and M. C. Adams, "Reactive chemical flow modeling applied to injection in the Coso EGS experiment," in *Proceedings of the 31st Workshop on Geothermal Reservoir Engineering*, Stanford University, Stanford, CA, January 30–February 1, 2006.

[134] V. Rabemanana, P. Durst, D. Bächler, F.-D. Vuataz and T. Kohl, "Geochemical modelling of the Soultz-sous-Forêts Hot Fractured Rock system," *Geothermics,* vol. 32, p. 645–653, 2003.

[135] J. D. Rimstidt and H. L. Barnes, "The kinetics of silica-water reactions," *Geochimica et Cosmochimica Acta,* vol. 44(11), pp. 1683–1699, 1980.

[136] I. Gaus, "Role and impact of CO_2–rock interactions during CO_2 storage in sedimentary rocks," *International Journal of Greenhouse Gas Control,* vol. 4(1), pp. 73–89, 2010.

[137] J. B. Randolph and M. O. Saar, "Combining geothermal energy capture with geologic carbon dioxide sequestration," *Geophysical Research Letters,* vol. 38(10), p. L10401, 2011.

[138] G. S. Kumar and A. Ghassemi, "Numerical modeling of non-isothermal quartz dissolution/precipitation in a coupled fracture–matrix system," *Geothermics,* vol. 34, p. 411–439, 2005.

[139] K. S. McLin, J. N. Moore, J. Hulen, J. R. Bowman and B. Berard, "Mineral characterization of scale deposits in injection wells; Coso and Salton Sea geothermal fields, CA," in *Proceedings of Thirty First Workshop on Geothermal Reservoir Engineering.* Stanford University, Stanford, CA, January 30–February 1, 2006.

[140] C. Steefel and M. Hu, "Reactive transport modeling of mineral precipitation and carbon trapping in discrete fracture networks," *Water Resources Research,* vol. 58(9), p. e2022WR032321, 2022.

[141] E. H. Rutter, "A Discussion on natural strain and geological structure - The kinetics of rock deformation by pressure solution," *Philosophical Transactions of the Royal Society of London. Series A, Mathematical and Physical Sciences,* vol. 283, p. 203–219, 1976.

[142] Y. Bernabé and B. Evans, "Numerical modelling of pressure solution deformation at axisymmetric asperities under normal load," *Geological Society, London, Special Publications,* vol. 284, p. 185–205, 2007.

[143] D. Elsworth and H. Yasuhara, "Mechanical and transport constitutive models for fractures subject to dissolution and precipitation," *International Journal for Numerical and Analytical Methods in Geomechanics,* vol. 34, p. 533–549, 2009.

[144] K.-B. Min, J. Rutqvist and D. Elsworth, "Chemically and mechanically mediated influences on the transport and mechanical characteristics of rock fractures," *International Journal of Rock Mechanics and Mining Sciences,* vol. 46, p. 80–89, 2009.

[145] S. Ogata, H. Yasuhara, N. Kinoshita, D.-S. Cheon and K. Kishida, "Modeling of coupled thermal-hydraulic-mechanical-chemical processes for predicting the evolution in permeability and reactive transport behavior within single rock fractures," *International Journal of Rock Mechanics and Mining Sciences,* vol. 107, p. 271–281, 2018.

[146] C. Rawal and A. Ghassemi, "A reactive thermo-poroelastic analysis of water injection into an enhanced geothermal reservoir," *Geothermics,* vol. 50, p. 10–23, 2014.

[147] M. An, F. Zhang, K.-B. Min, D. Elsworth, C. Marone and C. He, "The potential for low-grade metamorphism to facilitate fault instability in a geothermal reservoir," *Geophysical Research Letters,* vol. 48(11), p. e2021GL093552, 2021.

[148] K. Spokas, Y. Fang, J. P. Fitts, C. A. Peters and D. Elsworth, "Collapse of reacted fracture surface decreases permeability and frictional strength," *Journal of Geophysical Research: Solid Earth,* vol. 124, p. 12799–12811, 2019.

[149] G. Izadi and D. Elsworth, "Reservoir stimulation and induced seismicity: Roles of fluid pressure and thermal transients on reactivated fractured networks," *Geothermics,* vol. 51, p. 368–379, 2014.

[150] W. Rühaak, C.-D. Heldmanna and L. S. I. Peia, "Thermo-hydro-mechanical-chemical coupled modeling of a geothermally," *International Journal of Rock Mechanics and Mining Sciences,* vol. 100, p. 40–47, 2017.

[151] G. C. P. King, R. S. Stein and J. Lin, "Static stress changes and the triggering of earthquakes," *Bulletin of the Seismological Society of America,* vol. 84, p. 935–953, 1994.

[152] M. Schoenball, C. Baujard, T. Kohl and L. Dorbath, "The role of triggering by static stress transfer during geothermal reservoir stimulation," *Journal of Geophysical Research: Solid Earth,* vol. 117, B09307, 2012.

[153] Y. Okada, "Internal deformation due to shear and tensile faults in a half-space," *Bulletin of the Seismological Society of America,* vol. 82, p. 1018–1040, 1992.

[154] Y. Guglielmi, F. Cappa, J.-P. Avouac, P. Henry and D. Elsworth, "Seismicity triggered by fluid injection–induced aseismic slip," *Science,* vol. 348, p. 1224–1226, 2015.

[155] F. Cappa, M. M. Scuderi, C. Collettini, Y. Guglielmi and J.-P. Avouac, "Stabilization of fault slip by fluid injection in the laboratory and in situ," *Science Advances,* vol. 5, eaau4065, 2019.
[156] E. Majer, J. Nelson, A. Robertson-Tait, J. Savy and I. Wong, *Protocol for Addressing Induced Seismicity Associated with Enhanced Geothermal Systems, DOE/EE-0662,* Vol. 52, US Department of Energy, Washington, DC, 2012.
[157] D. Giardini, "Geothermal quake risks must be faced," *Nature,* vol. 462, p. 848–849, 2009.
[158] V. Vilarrasa, J. Carrera, S. Olivella, J. Rutqvist and L. Laloui, "Induced seismicity in geologic carbon storage," *Solid Earth,* vol. 10, p. 871–892, 2019.
[159] M. D. Zoback, "Managing the seismic risk posed by wastewater disposal" *Earth Magazine* vol. 57,p. 38–43, 2012.
[160] E. Gaucher, M. Schoenball, O. Heidbach, A. Zang, P. A. Fokker, J.-D. van Wees and T. Kohl, "Induced seismicity in geothermal reservoirs: A review of forecasting approaches," *Renewable and Sustainable Energy Reviews,* vol. 52, p. 1473–1490, 2015.
[161] F. Wenzel, A. Barth, C. Langenbruch and S. Shapiro, "Occurrence probability and earthquake size of post shut-in events in geothermal projects," in *Proceedings of the Workshop Induced Seismicity,* Luxembourg, 15–17 November 2010.
[162] A. Barth, F. Wenzel and C. Langenbruch, "Probability of earthquake occurrence and magnitude estimation in the post shut-in phase of geothermal projects," *Journal of Seismology,* vol. 17, p. 5–11, 2011.
[163] S. A. Shapiro, C. Dinske, C. Langenbruch and F. Wenzel, "Seismogenic index and magnitude probability of earthquakes induced during reservoir fluid stimulations," *The Leading Edge,* vol. 29, p. 304–309, 2010.
[164] B. Mena, S. Wiemer and C. Bachmann, "Building robust models to forecast the induced seismicity related to geothermal reservoir enhancement," *Bulletin of the Seismological Society of America,* vol. 103, p. 383–393, 2013.
[165] V. S. Gischig, D. Giardini, F. Amann, M. Hertrich, H. Krietsch, S. Loew, H. Maurer, L. Villiger, S. Wiemer, F. Bethmann, B. Brixel, J. Doetsch, N. G. Doonechaly, T. Driesner, N. Dutler, K. F. Evans, M. Jalali, D. Jordan, A. Kittilä, X. Ma, P. Meier, M. Nejati, A. Obermann, K. Plenkers, M. O. Saar, A. Shakas and B. Valley, "Hydraulic stimulation and fluid circulation experiments in underground laboratories: Stepping up the scale towards engineered geothermal systems," *Geomechanics for Energy and the Environment,* vol. 24, p. 100175, 2020.
[166] M. Galis, J. P. Ampuero, P. M. Mai and F. Cappa, "Induced seismicity provides insight into why earthquake ruptures stop," *Science Advances,* vol. 3(12), eaap7528, 2017.
[167] S. A. Shapiro, K.-H. Kim and J.-H. Ree, "Magnitude and nucleation time of the 2017 Pohang Earthquake point to its predictable artificial triggering," *Nature Communications,* vol. 12, p. 6397, 2021.
[168] P. Talwani and S. Acree, "Pore pressure diffusion and the mechanism of reservoir-induced seismicity," in *Earthquake Prediction,* Pure and applied geophysics, Eds. Shimazaki, K., Stuart, W. Birkhäuser, Basel, p. 947–965, 1985.
[169] S. A. Shapiro, E. Huenges and G. Borm, "Estimating the crust permeability from fluid-injection-induced seismic emission at the KTB site," *Geophysical Journal International,* vol. 131, p. F15–F18, 1997.
[170] S. A. Shapiro, P. Audigane and J.-J. Royer, "Large-scalein situpermeability tensor of rocks from induced microseismicity," *Geophysical Journal International,* vol. 137, p. 207–213, 1999.
[171] S. A. Shapiro, R. Patzig, E. Rothert and J. Rindschwentner, "Triggering of seismicity by pore-pressure perturbations: Permeability-related signatures of the phenomenon," *Pure and Applied Geophysics,* vol. 160, p. 1051–1066, 2003.

[172] M. Parotidis, S. A. Shapiro and E. Rothert, "Back front of seismicity induced after termination of borehole fluid injection," *Geophysical Research Letters,* vol. 31, L02612, 2004.
[173] S. A. Shapiro and C. Dinske, "Scaling of seismicity induced by nonlinear fluid-rock interaction," *Journal of Geophysical Research,* vol. 114, B09307, 2009.
[174] N. Hummel and T. M. Müller, "Microseismic signatures of non-linear pore-fluid pressure diffusion," *Geophysical Journal International,* vol. 179, p. 1558–1565, 2009.
[175] Y. Mukuhira, C. Dinske, H. Asanuma, T. Ito and M. O. Häring, "Pore pressure behavior at the shut-in phase and causality of large induced seismicity at Basel, Switzerland," *Journal of Geophysical Research: Solid Earth,* vol. 122, p. 411–435, 2017.
[176] L. Jing, "A review of techniques, advances and outstanding issues in numerical modelling for rock mechanics and rock engineering," *International Journal of Rock Mechanics and Mining Sciences,* vol. 40, p. 283–353, 2003.
[177] Q. Lei, J.-P. Latham and C.-F. Tsang, "The use of discrete fracture networks for modelling coupled geomechanical and hydrological behaviour of fractured rocks, *Computers and Geotechnics,* vol. 85, p. 151–176, 2017.
[178] J. Rutqvist, P. F. Dobson, J. Garcia, C. Hartline, P. Jeanne, C. M. Oldenburg, D. W. Vasco and M. Walters, "The northwest geysers EGS demonstration project, California: Pre-stimulation modeling and interpretation of the stimulation," *Mathematical Geosciences,* vol. 47, p. 3–29, 2015.
[179] F. Cappa and J. Rutqvist, "Modeling of coupled deformation and permeability evolution during fault reactivation induced by deep underground injection of CO_2," *International Journal of Greenhouse Gas Control,* vol. 5, p. 336–346, 2011.
[180] D. R. Faulkner, C. A. L. Jackson, R. J. Lunn, R. W. Schlische, Z. K. Shipton, C. A. J. Wibberley and M. O. Withjack, "A review of recent developments concerning the structure, mechanics and fluid flow properties of fault zones," *Journal of Structural Geology,* vol. 32, p. 1557–1575, 2010.
[181] Q. Lei, N. G. Doonechaly and C.-F. Tsang, "Modelling fluid injection-induced fracture activation, damage growth, seismicity occurrence and connectivity change in naturally fractured rocks," *International Journal of Rock Mechanics and Mining Sciences,* vol. 138, p. 104598, 2021.
[182] H. Yoo, S. Park, L. Xie, K.-I. Kim, K.-B. Min, J. Rutqvist and A. P. Rinaldi, "Hydro-mechanical modeling of the first and second hydraulic stimulations in a fractured geothermal reservoir in Pohang, South Korea," *Geothermics,* vol. 89, p. 101982, 2021.
[183] K. Kim, H. Yoo, S. Park, J. Yim, L. Xie, K. Min and J. Rutqvist, "Induced and triggered seismicity by immediate stress transfer and delayed fluid migration in a fractured geothermal reservoir at Pohang, South Korea," *International Journal of Rock Mechanics and Mining Sciences,* vol. 153, p. 105098, 2022.
[184] J. Rutqvist, A. P. Rinaldi, F. Cappa and G. J. Moridis, "Modeling of fault reactivation and induced seismicity during hydraulic fracturing of shale-gas reservoirs," *Journal of Petroleum Science and Engineering,* vol. 107, p. 31–44, 2013.
[185] M. W. McClure and R. N. Horne, "An investigation of stimulation mechanisms in Enhanced Geothermal Systems," *International Journal of Rock Mechanics and Mining Sciences,* vol. 72, p. 242–260, 2014.
[186] F. Parisio, V. Vilarrasa, W. Wang, O. Kolditz and T. Nagel, "The risks of long-term re-injection in supercritical geothermal systems," *Nature Communications,* vol. 10, 4391, 2019.
[187] P. Jeanne, J. Rutqvist, P. F. Dobson, M. Walters, C. Hartline and J. Garcia, "The impacts of mechanical stress transfers caused by hydromechanical and thermal processes on fault stability during hydraulic stimulation in a deep geothermal reservoir," *International Journal of Rock Mechanics and Mining Sciences,* vol. 72, p. 149–163, 2014.

[188] P. Segall and S. Lu, "Injection-induced seismicity: Poroelastic and earthquake nucleation effects," *Journal of Geophysical Research: Solid Earth,* vol. 120, p. 5082–5103, 2015.

[189] J. B. Altmann, B. I. R. Müller, T. M. Müller, O. Heidbach, M. R. P. Tingay and A. Weißhardt, "Pore pressure stress coupling in 3D and consequences for reservoir stress states and fault reactivation," *Geothermics,* vol. 52, p. 195–205, 2014.

[190] K. W. Chang and P. Segall, "Injection-induced seismicity on basement faults including poroelastic stressing," *Journal of Geophysical Research: Solid Earth,* vol. 121, p. 2708–2726, 2016.

[191] M. Cacace, H. Hofmann and S. A. Shapiro, "Projecting seismicity induced by complex alterations of underground stresses with applications to geothermal systems," *Scientific Reports,* vol. 11, 23560, 2021.

[192] F. Catalli, M.-A. Meier and S. Wiemer, "The role of Coulomb stress changes for injection-induced seismicity: The Basel enhanced geothermal system," *Geophysical Research Letters,* vol. 40, p. 72–77, 2013.

[193] I. W. Yeo, M. R. M. Brown, S. Ge and K. K. Lee, "Causal mechanism of injection-induced earthquakes through the Mw 5.5 Pohang earthquake case study," *Nature Communications,* vol. 11, p. 2614, 2020.

[194] V. Vilarrasa, S. De Simone, J. Carrera and A. Villaseñor, "Unraveling the causes of the seismicity induced by underground gas storage at Castor, Spain," *Geophysical Research Letters,* vol. 48, e2020GL092038, 2021.

[195] S. Wei, J.-P. Avouac, K. W. Hudnut, A. Donnellan, J. W. Parker, R. W. Graves, D. Helmberger, E. Fielding, Z. Liu, F. Cappa and M. Eneva, "The 2012 Brawley swarm triggered by injection-induced aseismic slip," *Earth and Planetary Science Letters,* vol. 422, p. 115–125, 2015.

[196] L. Duboeuf, L. D. Barros, F. Cappa, Y. Guglielmi, A. Deschamps and S. Seguy, "Aseismic motions drive a sparse seismicity during fluid injections into a fractured zone in a carbonate reservoir," *Journal of Geophysical Research: Solid Earth,* vol. 122, p. 8285–8304, 2017.

[197] T. S. Eyre, D. W. Eaton, D. I. Garagash, M. Zecevic, M. Venieri, R. Weir and D. C. Lawton, "The role of aseismic slip in hydraulic fracturing–induced seismicity," *Science Advances,* vol. 5, eaav7172, 2019.

[198] J. Rutqvist, A. P. Rinaldi, F. Cappa, P. Jeanne, A. Mazzoldi, L. Urpi, Y. Guglielmi and V. Vilarrasa, "Fault activation and induced seismicity in geological carbon storage – Lessons learned from recent modeling studies," *Journal of Rock Mechanics and Geotechnical Engineering,* vol. 8, p. 789–804, 2016.

[199] Q. Lei, X. Wang, K.-B. Min and J. Rutqvist, "Interactive roles of geometrical distribution and geomechanical deformation of fracture networks in fluid flow through fractured geological media," *Journal of Rock Mechanics and Geotechnical Engineering,* vol. 12, p. 780–792, 2020.

[200] M. Alghannam and R. Juanes, "Understanding rate effects in injection-induced earthquakes," *Nature Communications,* vol. 11, p. 3053, 2020.

[201] D. Dempsey and J. Riffault, "Response of induced seismicity to injection rate reduction: Models of delay, decay, quiescence, recovery, and Oklahoma," *Water Resources Research,* vol. 55, p. 656–681, 2019.

[202] E. Rothert and S. A. Shapiro, "Microseismic monitoring of borehole fluid injections: Data modeling and inversion for hydraulic properties of rocks," *Geophysics,* vol. 68, p. 685–689, 2003.

[203] M. Parotidis and S. A. Shapiro, "A statistical model for the seismicity rate of fluid-injection-induced earthquakes," *Geophysical Research Letters,* vol. 31, L17609, 2004.

[204] S. A. Shapiro, C. Dinske and J. Kummerow, "Probability of a given-magnitude earthquake induced by a fluid injection," *Geophysical Research Letters,* vol. 34, L22314, 2007.
[205] C. Langenbruch and S. A. Shapiro, "Decay rate of fluid-induced seismicity after termination of reservoir stimulations," *Geophysics,* vol. 75, p. MA53–MA62, 2010.
[206] C. E. Bachmann, S. Wiemer, B. P. Goertz-Allmann and J. Woessner, "Influence of pore-pressure on the event-size distribution of induced earthquakes," *Geophysical Research Letters,* vol. 39, L09302, 2012.
[207] B. P. Goertz-Allmann and S. Wiemer, "Geomechanical modeling of induced seismicity source parameters and implications for seismic hazard assessment," *Geophysics,* vol. 78, p. KS25–KS39, 2013.
[208] V. S. Gischig and S. Wiemer, "A stochastic model for induced seismicity based on non-linear pressure diffusion and irreversible permeability enhancement," *Geophysical Journal International,* vol. 194, p. 1229–1249, 2013.
[209] A. P. Rinaldi and M. Nespoli, "TOUGH2-seed: A coupled fluid flow and mechanical-stochastic approach to model injection-induced seismicity," *Computers & Geosciences,* vol. 108, p. 86–97, 2017.
[210] F. Amann, V. Gischig, K. Evans, J. Doetsch, R. Jalali, B. Valley, H. Krietsch, N. Dutler, L. Villiger, B. Brixel, M. Klepikova, A. Kittilä, C. Madonna, S. Wiemer, M. O. Saar, S. Loew, T. Driesner, H. Maurer and D. Giardini, "The seismo-hydro-mechanical behaviour during deep geothermal reservoir stimulations: Open questions tackled in a decameter-scale in-situ stimulation experiment," *Solid Earth,* vol. 9(1), pp. 115–137, 2017.

Chapter 7

Hydraulic stimulation of tight geothermal reservoirs

Sri Kalyan Tangirala
Indian Institute of Technology (Indian School of Mines), Dhanbad
Spanish National Research Council (IMEDEA, CSIC-UIB)

Victor Vilarrasa
Spanish National Research Council (IMEDEA, CSIC-UIB)

Francesco Parisio
Spanish National Research Council (IMEDEA, CSIC-UIB)
Spanish National Research Council (IDAEA-CSIC)

CONTENTS

7.1 INTRODUCTION: BACKGROUND AND MOTIVATION

In enhanced geothermal systems (EGS), hydraulic stimulation aims at increasing the permeability of the crystalline basement to circulate fluids at a rate that is sufficient to produce electricity from geothermal energy (Abe and Horne, 2023). The physical and economic viability of an EGS project strongly depend on the possibility of enhancing the permeability of the reservoir either by propagating new hydraulic fractures or by improving the permeability of existing ones by inducing shear dilation (Vilarrasa et al., 2011; Parisio and Yoshioka, 2020). Most of the EGSs are typically developed in the crystalline basement (Brown et al., 2012), with high amounts of heat trapped in them. Several field studies involving granitic EGSs have been carried out around the world since the 1970s, e.g., Fenton Hill in the US, Rosemanowes in the U.K., Hijiori in Japan, Cooper Basin in Australia, Soultz-sous-Forêt in France and Basel in Switzerland (reviewed by McClure and Horne, 2014). The typical challenges that occur during EGS stimulation include the inability to sufficiently enhance the permeability of the existing fractures, the difficulties in joining the injection and production wells with a network of permeable fractures (Lepillier et al., 2020), the so-called thermal short-circuiting (Ghassemi, 2012), and induced seismicity (Majer et al., 2007). Induced seismicity proved to be a major roadblock to EGS projects following the events at Basel, Switzerland (Häring et al., 2008)

DOI: 10.1201/9781003271475-7

and Pohang, Korea Republic (Ellsworth et al., 2019). In an attempt to maximize permeability enhancement and minimize the magnitude, *M*, and frequency of the induced earthquakes, a wide range of stimulation protocols have been proposed and employed (McClure and Horne, 2014). Despite the relative abundance, none of the protocols has delivered the expected results: often, the increment of permeability was insufficient (Xie et al., 2015) and/or the magnitude and frequency of the induced earthquakes were above acceptable levels (Giardini, 2009; Ellsworth et al., 2019). As an example, traffic light systems (TLS) have been devised in an attempt to limit the magnitude and frequency of induced earthquakes during the stimulation stage (Verdon and Bommer, 2021). TLS, however, did not prevent the M3.4 Basel earthquake in December 2004 (Häring et al., 2008) nor the M5.5 Pohang earthquake in November 2017 (Ellsworth et al., 2019).

At Basel, the reactivation of the network of natural fractures occurred during hydro-shearing stimulation with the step-rate injection scheme (Mukuhira et al., 2013). The fractures were reactivated shortly after the beginning of the hydraulic stimulation, and the magnitude and frequency of the induced events progressively increased during the six days of injection (Deichmann et al., 2014). The injection was stopped after an event with $M > 2$, but a M3.4 earthquake occurred a few hours after shut-in (Mukuhira et al., 2017). The Basel EGS project was eventually cancelled as a result of the negative impacts caused by induced seismicity (Andrés et al., 2019). It is a common occurrence in induced seismicity that the magnitude of post-injection events is up to three orders of magnitude larger than the events occurring during the injection phase in terms of seismic moment. Although the cause of the larger-magnitude post-injection events is likely to be related to poromechanical stress induced by pressure diffusion and shear-slip stress transfer caused by induced microseismic events (Kim and Makhnenko, 2020), we are still far from an accurate quantification of the driving mechanisms (Segall and Lu, 2015; De Simone et al., 2017).

At Pohang, the cyclic-injection protocol was employed during stimulation. A total of five different stimulation stages took place in two separate wells called PX-1 and PX-2. The third and fifth stimulations, both performed at PX-2, induced earthquakes of magnitude 3.1 and 5.5, respectively, in a low-permeable fault located between the two wells (Kim et al., 2018). The M5.5 earthquake, which represents the largest earthquake ever induced by the hydraulic stimulation of an EGS project, also occurred in the post-injection stage and led to project cancellation. The delay between the shut-in and the earthquake was two months at Pohang, much longer than at Basel (Yeo et al., 2020). For the given injected volume, the magnitude was significantly higher than its theoretical maximum as predicted by McGarr's relationship (McGarr, 2014), which is used as the basis for the TLS control on some occasions. The formulations that bound the maximum magnitude to the injected volume have several shortcomings since they do not take into account specific tectonic and geological settings, hydro-geological parameters or injection protocols.

Investigation of the coupled physical processes that govern the response of tight rock to high-pressure injection during hydraulic stimulation is a necessary step to safely conduct EGS operations (Cueto-Felgueroso et al., 2017; Pampillon et al., 2018; Zareidarmiyan et al., 2021). An effective design phase will not only minimize the risks of damage, disruption and potential harm to human life but will also minimize the financial risks connected to the project. A thorough characterization accompanied by extensive numerical modelling in the design phase of injection protocols will increase the chances of striking an optimum balance between maximizing productivity (enhancing permeability) and minimizing risks (induced seismicity). This study demonstrates the importance of considering site- and project-specific information in the design of EGS stimulation. We overcome the limitations of the volume-magnitude scaling laws by investigating the impact of the injection protocol on permeability enhancement and induced seismicity. We simulate a series of cases in which the same amount of water is injected into the reservoir but with different injection protocols (scheme and injection rate) and study the response in terms of permeability enhancement and slip induced in the fracture. The constitutive model of the fracture integrates a slip weakening law and dilation phenomena (Brace et al., 1966). We compare the permeability enhancement and the slip induced along the fracture to assess the effectiveness of the stimulation protocols and find that the cyclic-injection scheme enhances the permeability of a larger area of the fracture compared to the other injection schemes, but at the expense of experiencing a larger slip, which may lead to more induced seismicity than with the other schemes. The models with step-rate injection induce the least amount of slip in the fracture and induce seismicity, as well as the least permeability enhancement.

7.2 METHODS

We run bi-dimensional (2D) hydromechanical (HM) simulations on a square-shaped granitic reservoir of side length 500 m, where water is injected at the centre of a pre-existing fracture/fault that forms an angle of 70° with the minimum principal stress (Figure 7.1a). The 2D model represents a horizontal plane intersected by a vertical fracture, so gravity is neglected in the simulations. The fracture is modelled as a 10-cm-thick continuum medium. We use three different injection schemes: constant rate, step rate and cyclic injection. Additionally, each scheme is composed of three different injection rates: 0.1, 0.2 and 0.3 kg/s/m applied to the constant and cyclic schemes. There are five cycles in each cyclic scheme, where each cycle comprises an injection interval followed by a no-injection interval of equal duration. The step-rate injection scheme is composed of five stages of increasing injection rate, with the different values of flow rates corresponding to the latest stage. To observe the effects of dilatancy on permeability enhancement, we use four values of the dilatancy angle – 0.1°, 2°, 4° and 10° – for every injection scheme and injection rate. We run a total of 36 simulations – constant-rate scheme with a 0.1 kg/s/m rate for 0.1° dilatancy angle being one example. The amount of water injected into the reservoir is 270 kg/m in all models.

The coupled HM simulations are solved monolithically. For the hydraulic problem, mass conservation of water is given by

$$\frac{\varphi}{K_f}\frac{\partial p}{\partial t} + \nabla \cdot \frac{du}{dt} + \frac{1}{\rho}\nabla \cdot (\rho q) = f_w, \tag{7.1}$$

where φ is the porosity, $\frac{1}{K_f}$ is the water compressibility, p is the pressure, u is the solid displacement vector, t is the time, ρ is the water density, q is the water flux and f_w is the sink/source term. For the mechanical problem, ignoring the inertial terms, the momentum balance equation simplifies into the equilibrium of stresses

$$\nabla \cdot \sigma + f = 0, \tag{7.2}$$

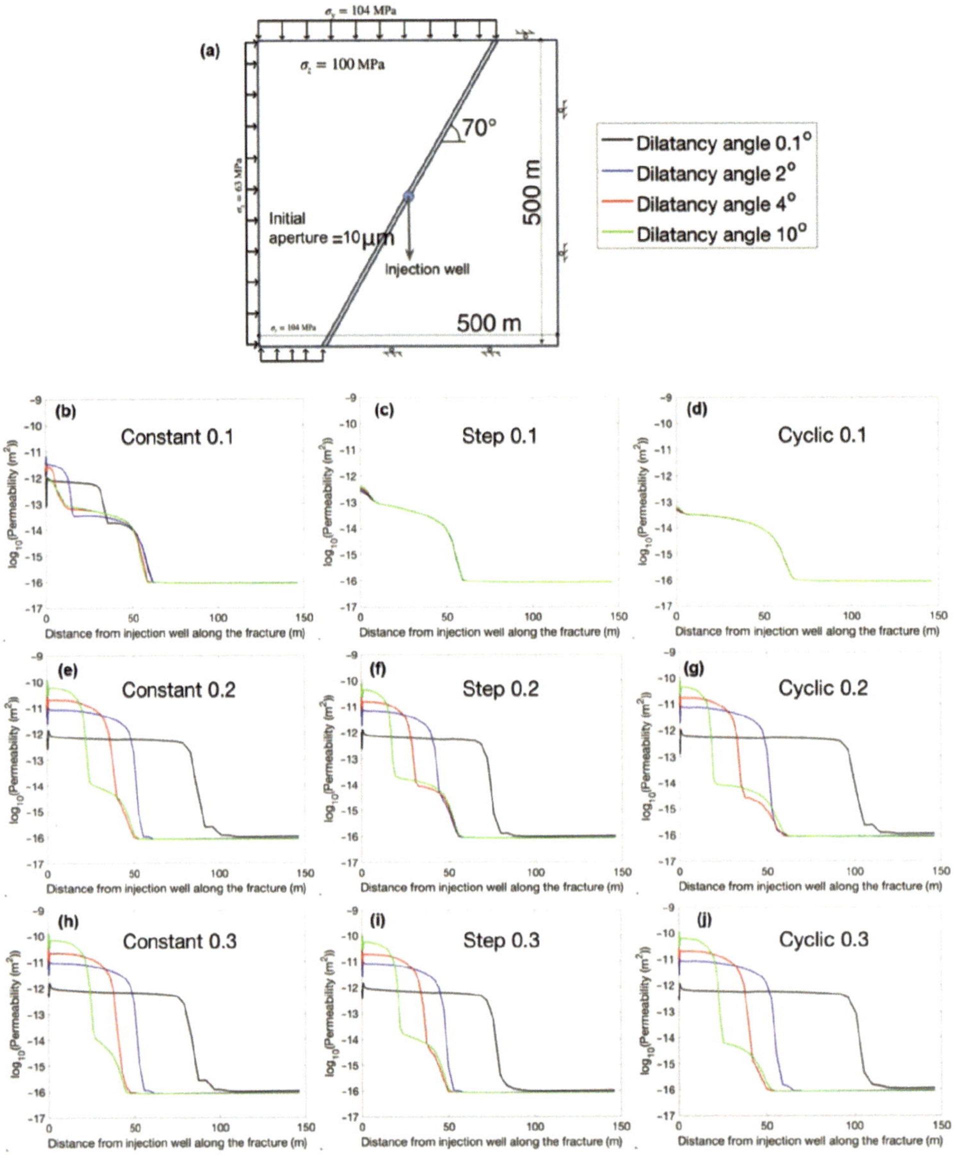

Figure 7.1 (a) Geometry of the model and legend for plots (b–j), which represent the permeability as a function of the dilatancy angle along the fracture at the end of the stimulation for various injection schemes (constant (b, e and h), step (c, f and i) and cyclic (d, g and j)) and flow rates (0.1 (b–d), 0.2 (e–g) and 0.3 (h–j) kg/s/m) as mentioned in the plots.

Table 7.1 Material properties of rock matrix the fracture

Material properties	*Rock matrix*	*Fracture*
Permeability (m^2)	$2.5 \cdot 10^{-18}$	$8.33 \cdot 10^{-17}$
Porosity (–)	0.005	0.0001
Young's modulus (GPa)	60	2.8
Poisson's ratio (–)	0.3	0.44
Biot's coefficient (–)	0.5	1
1/Viscosity (s^{-1} MPa^{-1})	–	$1 \cdot 10^{-4}$
Maximum cohesion (MPa)	–	0.1
Residual cohesion (MPa)	–	0.1
Maximum friction angle (°)	–	36
Residual friction angle (°)	–	24
Dilatancy angle (°)	–	0.1, 2, 4 and 10
Critical value of softening parameter (–)	–	0.0185

where σ is the stress tensor and f is the vector of body forces. Darcy's law is used to compute the flow of water in the system and writes

$$\boldsymbol{q} = -\frac{k}{\mu}(\nabla p + \rho g \nabla z), \tag{7.3}$$

where k is the permeability, μ is the water viscosity and g is the gravity.

The rock matrix and the fracture are considered as two different materials with distinct properties listed in Table 7.1. Fracture permeability varies as a function of aperture, b, following the cubic law (Olivella and Alonso, 2008)

$$k = k_{matrix} + \frac{b^3}{12a}, \tag{7.4}$$

$$b = b_0 + \Delta b, \tag{7.5}$$

$$\Delta b = a\Delta\epsilon = a(\epsilon - \epsilon_o), \quad \text{for } \epsilon > \epsilon_o, \tag{7.6}$$

where k_{matrix} is the intrinsic permeability of the porous material filling the fracture, a is the spacing of the fractures, and ϵ and ϵ_o are the volumetric strain and reference volumetric strain, respectively. The initial aperture (b_0) is 10 μm and the spacing between the fractures (a) is 10 cm, i.e., only one fracture is present in the continuum medium representing the fracture.

The constitutive behaviour of fractures is governed by a plastic model that includes dilatancy and strain softening. The yield function is

$$F = \sigma_m \cdot \sin\phi(\eta) + \left(\cos\theta - \frac{1}{\sqrt{3}}\sin\theta \cdot \sin\phi(\eta)\right) \cdot \sqrt{J_2} - c(\eta) \cdot \cos\phi(\eta) \tag{7.7}$$

and the plastic potential

$$G = \alpha \cdot \sigma_m \cdot \sin\psi + \left(\cos\theta - \frac{1}{\sqrt{3}}\sin\theta \cdot \sin\psi\right) \cdot \sqrt{J_2} - c(\eta) \cdot \cos\psi, \tag{7.8}$$

while the loading-unloading conditions are

$$\Phi(F) = F^m \text{ for } F \geq 0 \quad \text{and} \quad \Phi(F) = 0 \text{ for } F < 0. \tag{7.9}$$

The previous equations express the plastic behaviour as a combination of invariants of the stress tensor, i.e.,

$$\sigma_m = (\sigma_{xx} + \sigma_{yy} + \sigma_{zz}) / 3;$$

$$J_2 = \frac{1}{6}\sqrt{(\sigma_{xx} - \sigma_{yy})^2 + (\sigma_{yy} - \sigma_{zz})^2 + (\sigma_{zz} - \sigma_{xx})^2 + 6(\tau_{xy}^2 + \tau_{yz}^2 + \tau_{zx}^2)};$$

$$\theta = \frac{1}{3}\sin^{-1}\left(-\frac{3\sqrt{3}J_3}{2\sqrt{J_2}^3}\right);$$

$$J_3 = S_x S_y S_z + 2\tau_{xy}\tau_{yz}\tau_{zx} - S_x\tau_{yz}^2 - S_y\tau_{xz}^2 - S_z\tau_{xy}^2.$$

In the plastic model, ψ is the dilatancy angle, c the cohesion, and the friction angle ϕ

$$\phi(\eta) = \left\{\phi^{peak},\ \eta \leq 0\ ; \eta^{\ peak} + \left(\frac{\phi^{res} - \phi^{peak}}{\eta^*}\right) \cdot \eta,\ 0 \leq \eta \leq \eta^*;\ \eta^{res},\ \eta^* \leq \eta\right\}, \tag{7.10}$$

where

$$\eta = \sqrt{\frac{3}{2} \cdot \left[\left(\epsilon_{xx}^p - \epsilon_m^p\right)^2 + \left(\epsilon_{yy}^p - \epsilon_m^p\right)^2 + \left(\epsilon_{zz}^p - \epsilon_m^p\right)^2 + \left(\frac{1}{2} \cdot \gamma_{xy}^p\right)^2 + \left(\frac{1}{2} \cdot \gamma_{yz}^p\right)^2 + \left(\frac{1}{2} \cdot \gamma_{zx}^p\right)^2\right]} \tag{7.11}$$

and

$$\epsilon_m^p = \frac{1}{3} \cdot \left(\epsilon_{xx}^p + \epsilon_{yy}^p + \epsilon_{zz}^p\right);$$

α is a parameter for plastic potential, ϵ^p is the plastic strain tensor and γ is the shear component of the plastic strain tensor.

The rock matrix in the reservoir geometry (Figure 7.1a) is ideally divided into two sectors separated by the fracture – one on the left-hand side of the fracture and the other on the right-hand side of the fracture. Positive x is towards the right and positive y is upwards in Figure 7.1a. To allow for large shear slip of the fracture upon reactivation, the boundaries of the right-hand side of the fracture are fixed, while constant stress is applied to the boundaries of the left-hand side. The minimum horizontal stress (σ_x) is 63 MPa, the maximum horizontal stress (σ_y) is 104 MPa and the vertical stress (σ_z) is 100 MPa. A constant pore pressure of 40 MPa is applied over the entire domain.

The finite element code CODE_BRIGHT (Olivella et al., 1994, 1996) is used to numerically simulate the fully coupled HM models. The mesh is composed of a total of 3,625 quadratic elements, being fine near the injection well and getting gradually coarser away from it along the fracture. The size of the elements is 3.33×10^{-2} m $\times$ 5.52×10^{-2}m near the well, 3.33×10^{-2} m $\times$ 6.68 m at the edges of the fracture, and 56.1 m $\times$ 50.0 m along the boundaries of the model.

7.3 RESULTS

The injection protocol affects the HM response of the fracture to hydraulic stimulation. Despite the same volume of water being injected in all cases, pore pressure build-up and the induced poromechanical stresses differ in each case. The difference is highlighted for the injection rate of 0.1 kg/s/m, which leads to purely elastic deformation in the cyclic-injection scheme, a slight fracture reactivation close to the injection well in the step-rate scheme, and a full fracture reactivation in the constant-rate scheme (compare Figure 7.1b–d). As the injection rate is increased, the maximum pore pressure increases as well, leading to fracture reactivation in all cases (Figure 7.1e–j). The pressure build-up does not scale linearly with the injection rate because fracture permeability is enhanced with the cube of the fracture aperture (Eq. 7.4). The fracture aperture increases during injection following two distinct mechanisms: pressure build-up and slip-induced dilatancy (Yeo et al., 1998). The larger the dilatancy angle, the larger the permeability enhancement, and thus, the lower the resulting pressure build-up. As a result, the length of the fracture that becomes reactivated decreases as the dilatancy angle increases (Figure 7.1b,e–j). For the largest dilatancy angle, i.e., 10°, a high-permeability region is created close to the injection well, but the permeability enhancement occurs in a relatively small area. In contrast, very low dilatancy angles, i.e., 0.1°, lead to a larger area undergoing permeability enhancement, but with the resulting permeability being two orders of magnitude lower than for the largest dilatancy angle. The magnitude of the permeability enhancement is similar for the three injection schemes, but the extent of the stimulated region depends on the scheme. The cyclic-injection scheme, closely followed by the constant-rate scheme, produces the largest fracture length in which permeability is enhanced at the end of the stimulation. The step-rate scheme yields the smallest stimulated fracture length, especially for low dilatancy angles. The pressure changes that occur within the fracture as a result of slip-induced dilatancy are accompanied by pore pressure changes in the surrounding rock matrix.

Shear slip induces additional pore pressure changes resulting from an undrained response of the rock matrix to the deformation caused by fracture reactivation (Vilarrasa et al., 2021). The difference in the pore pressure distribution between a stimulation that only induces elastic strain and another one that reactivates the fracture is displayed in Figure 7.2. The pore pressure distribution is purely diffusive when the fracture behaves elastically (Figure 7.2a). The poromechanical response to fracture pressurization is observed at the tips of the pressure diffusion front, where a pressure drop occurs because of deformation-induced pressure changes analogous to the Noordbergum effect (Hsieh, 1996). Once the fracture reaches failure conditions, it undergoes irreversible shear slip, which contracts the rock in front of the slip direction and expands the rock behind the slipped area. As a result, pore pressure increases where the rock is contracted and decreases where it is expanded (Figure 7.2b). Such pore pressure changes, which have a similar magnitude to the ones induced by injection within the fracture, are transient and dissipate after several days for the rock properties considered here. The dissipation time is inversely proportional to the rock permeability, and, thus, it could give rise to delays of weeks or months in a very low-permeability matrix.

Shear slip also induces changes in the shear stress at and around the fracture (Figure 7.3). The slipped area undergoes a shear stress drop that inhibits further

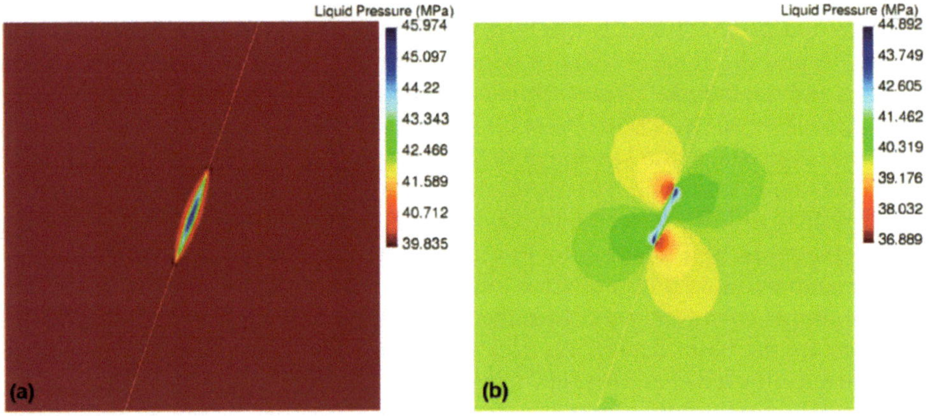

Figure 7.2 Pore pressure distribution at the end of stimulation for the cyclic-injection scheme with injection rates of (a) 0.1 kg/s/m and (b) 0.3 kg/s/m. Pore pressure distribution is indicative of elastic deformation in (a) and irreversible fracture reactivation in (b).

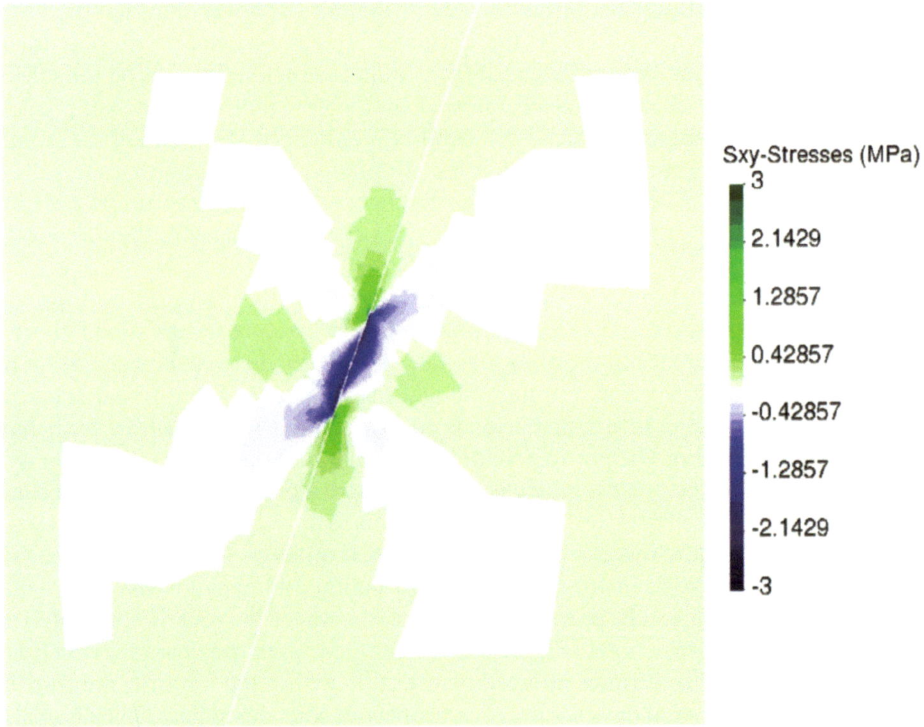

Figure 7.3 Shear stress distribution over the reservoir at the end of the cyclic-injection scheme with an injection rate of 0.3 kg/s/m.

reactivation. In contrast, the trailing edges of the slipped area experience an increase in shear stress that brings the stress state closer to failure conditions. However, stability is maintained at the edges shortly after slip because the deformation-induced pore pressure changes yield a pressure drop in these regions (compare Figures 7.2b and 7.3). Subsequently, as deformation-induced pore pressure changes dissipate, failure conditions may be reached because the shear stress changes remain (Vilarrasa et al., 2021). Such post-injection reactivation is observed in the cyclic-injection protocols, in which the stimulated region of the fracture extends at the tips during the no-injection intervals. This reactivation happens due to pore pressure diffusion towards the unstimulated regions of the fracture where shear stress has increased. At the end of the cyclic-stimulation protocol at a rate of 0.3 kg/s/m, the tips of the reactivated regions of the fracture are still weak (Figure 7.3), and the shear stress is sufficient to induce slip in the post-injection interval (Kamali and Ghassemi, 2016).

The amount of slip induced in the fracture during stimulation operations varies with the type of injection protocol and the injection rate used. The cyclic-injection protocol with a rate of 0.1 kg/s/m induces negligible slip (purely elastic) until the end of injection (Figure 7.4c). In the rest of the injection rates, the step-rate injection protocols

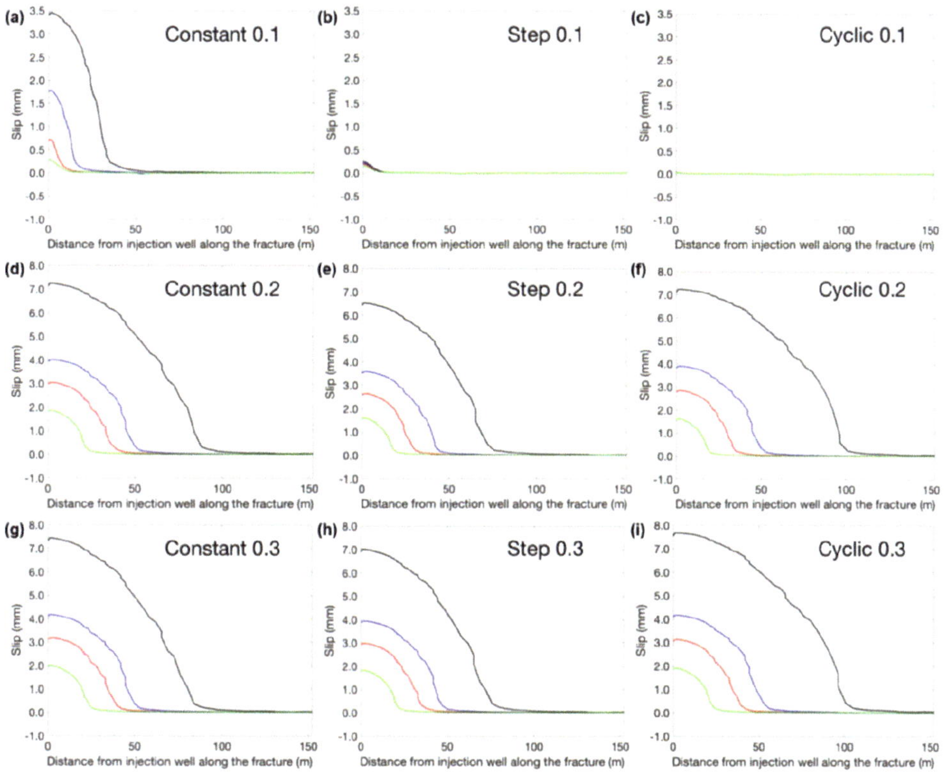

Figure 7.4 The slip along the fracture as a function of dilatancy for constant rate (a, d and g), step rate (b, e and h) and cyclic injection (c, f and i) at a rate of 0.1 (a, b and c), 0.2 (d, e and f) and 0.3 kg/s/m (g, h and i) at the end of the stimulations (the legend is the same as in Figure 7.1a).

induce less slip than the constant and cyclic protocols (Figure 7.4d–i). In step-rate injection, injection starts at a rate that is 20% of the rate injected in the constant and cyclic-stimulation protocols and gradually increases until it reaches the maximum rate, which is equal to the rate of injection in the constant and cyclic protocols. As a result, the pore pressure front progresses slower in the models with step-rate injection, accumulating less slip before the final injection rate is reached. During the final step, the amount of slip induced is considerable (Figure 7.5e and h), but the final cumulated slip at the end is lower than in the other protocols (Figure 7.4e and h).

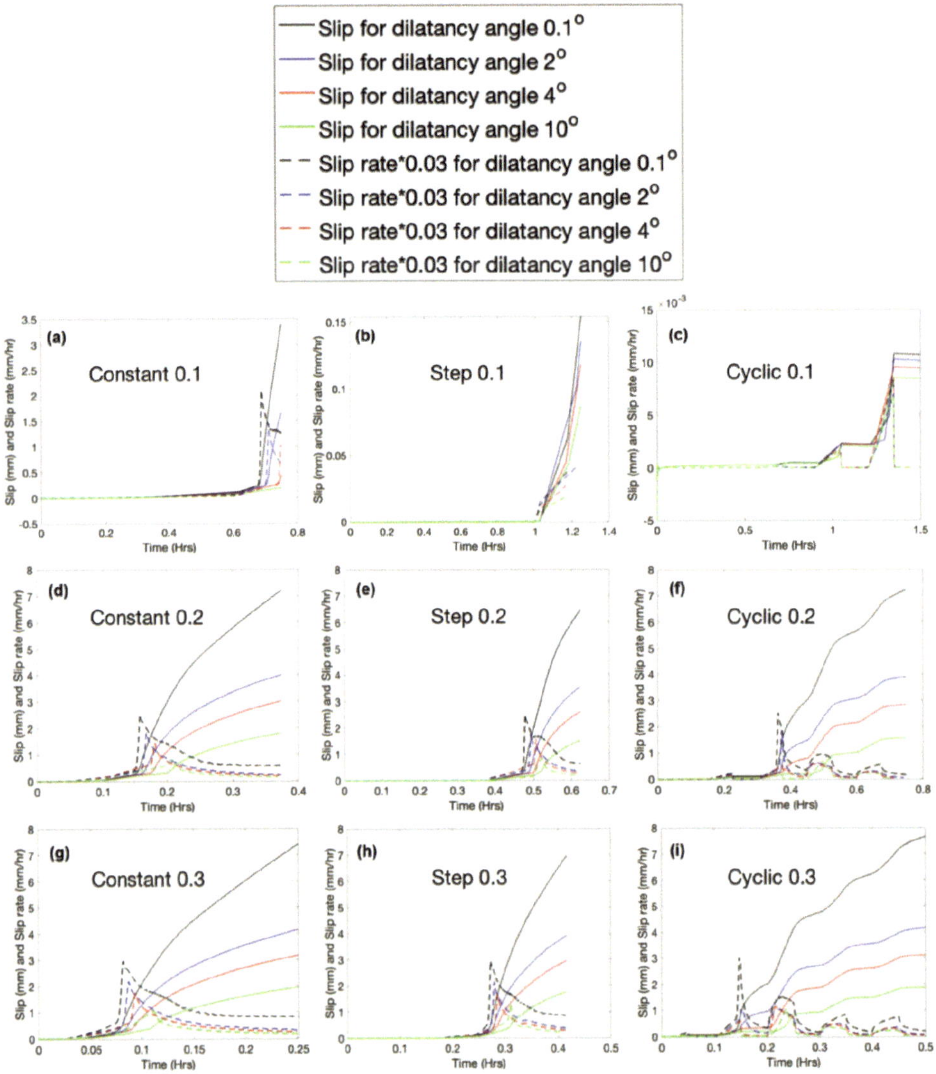

Figure 7.5 The slip and slip rate evolution as a function of dilatancy for constant rate (a, d and g), step rate (b, e and h) and cyclic injection (c, f and i) at a rate of 0.1 (a, b and c), 0.2 (d, e and f) and 0.3 kg/s/m (g, h and i).

The cyclic-injection protocols induce slip on a larger portion of the fracture than the other protocols due to the slip induced by the pore pressure diffusion at the edges of the reactivated region of the fracture in its alternate no-injection intervals (Figure 7.4f and i). The amount of injected water required to cause a significant slip in the fracture is larger than in the constant-rate protocol for an injection rate of 0.2 kg/s/m (compare Figure 7.5d and f), but they become similar as the injection rate increases to 0.3 kg/s/m (compare Figure 7.5g and i). This delay is caused by the pore pressure reduction that occurs during the no-injection intervals, retarding the time at which the critical pressure for reactivation is reached. Once the fracture is reactivated for the first time, a significant shear slip occurs at the beginning of each cycle. The constant-rate injection protocols lead to the earliest fracture reactivation (Figure 7.5a, d and g). Despite the differences in the time of reactivation and differences in the slip magnitude, the maximum slip rate, which coincides with the onset of shear slip, is very similar for the three protocols at injection rates higher than 0.1 kg/s/m (Figure 7.5).

7.4 DISCUSSION

We have conducted a numerical study to understand the mechanisms that differentiate the various stimulation protocols. Our results show that both permeability enhancement and injection-induced slip are influenced by the stimulation protocol. We observe a marked difference in the response between the injection rates of 0.1 and 0.2 kg/s/m for all the tested protocols (Figures 7.1b–g and 7.4a–f). Simulation results suggest the existence of a lower limit of injection rate for which fracture stimulation will not guarantee enough permeability. When the injection rate is lower than a given threshold (which intrinsically depends on the hydromechanical properties of the rock and fracture, fracture orientation with respect to the principal stresses and in-situ initial stress state), the injection-induced pressure build-up might not be sufficient to reactivate a fracture, which instead behaves in a purely elastic manner (Figures 7.1d and 7.4c). Above that threshold, the fracture reactivates at a given point, accumulating shear slip and enhancing permeability as a result of dilatancy.

Dilatancy enhances the permeability of the sheared region by one to four orders of magnitude (Figure 7.1). Low dilatancy angles give rise to a relatively low-permeability enhancement, but the length of the fracture with enhanced permeability is larger than for high dilatancy angles. As a result, the injection pressure for a given flow rate may increase, but so may the chances of establishing a hydraulic connection between the two wells of a doublet. Taking into account that dilatancy decreases with normal stress, the dilatancy angle becomes low at the typical depths of EGS (Xie and Min, 2021). For a given value of the dilatancy angle and injection rate (0.2 or 0.3 kg/s/m), the maximum achieved permeability is independent of the injection protocols. Instead, the protocol affects the extent of the fracture that becomes stimulated, which is the maximum for the cyclic injection (Figures 7.1e–7.4j). Yet, fracture reactivation may continue after shut-in in the constant and step-rate injection protocols, which may eventually lead to similar stimulated fracture lengths. The post-injection reactivation is caused by a combination of increased slip-induced shear stress and a reduction of the pore pressure, which drops during injection as a result of deformation-induced pore pressure changes (Vilarrasa et al., 2021).

Simulation results show that the cyclic-injection protocols also induce the largest amount of slip on the fracture (Figure 7.4f and i). The largest slip observed in the cyclic-injection protocols seems to suggest a higher potential for inducing larger earthquakes than the other schemes. However, our quasi-static analysis cannot differentiate between a-seismic and co-seismic dynamic slip, and the final size of runaway rupture requires full dynamic simulations. The slip rate is very similar for all protocols for a given injection rate, which hints at a similar timing of fault rupture. In our simulations, the maximum slip velocity is four orders of magnitude lower than the one for the onset of dynamic rupture, i.e., 0.1 m/s (Rubin and Ampuero, 2005), which could indicate that the system has not reached favourable conditions for runaway ruptures. Our results have shown that the influence of the adopted stimulation protocol is marginal. We have found no evidence that the cyclic-injection protocol is an effective strategy to control induced seismicity (Zang et al., 2019; Niemz et al., 2020). This result seems to be supported by the failure of the cyclic-stimulation protocol to prevent the large event that occurred in Pohang (Ellsworth et al., 2019).

7.5 CONCLUSIONS

We have studied the efficiency of three protocols of hydraulic stimulation in a single pre-existing fracture. The injection rate should be sufficiently high so that the fracture is reactivated and irreversible shear slip occurs, enhancing fracture permeability due to dilation. Permeability enhancement increases with the dilatancy angle but is concomitant with a decreasing stimulated fracture length. The cyclic-injection protocol enhances the permeability of a larger area of the fracture than the other protocols, and the step-rate injection protocol induces the least amount of slip in the fracture. Nonetheless, the differences remain within the same order of magnitude, and, thus, no significant differences between the hydraulic stimulation protocols exist. It is likely that de-risking induced seismicity will not be achieved with the current stimulation protocols that have been tested.

ACKNOWLEDGEMENTS

V.V. acknowledges funding from the European Research Council (ERC) under the European Union's Horizon 2020 Research and Innovation Program through the Starting Grant GEoREST (www.georest.eu) under Grant agreement No. 801809. F.P. and V.V. acknowledge funding from the European Union's Horizon 2020 Research and Innovation Programme through the Marie Sklodowska-Curie Action ARMISTICE under Grant Agreement No. 882733. IDAEA-CSIC is a Centre of Excellence Severo Ochoa (Spanish Ministry of Science and Innovation Grant CEX2018-000794-S funded by MCIN/AEI/ 10.13039/501100011033). IMEDEA is an accredited "Maria de Maeztu Excellence Unit" (Grant CEX2021-001198, funded by MCIN/AEI/10.13039/501100011033).

REFERENCES

Abe, A. and Horne, R. N., Investigating fracture network creation and stimulation mechanism of EGS reservoirs. *Geothermics*, *107*, 102606 (2023).

Andrés, S., Santillán, D., Mosquera, J.C., Cueto-Felgueroso, L., Thermo-poroelastic analysis of induced seismicity at the Basel enhanced geothermal system. *Sustainability*, *11*, 6904 (2019).

Brace, W.F., Paulding, E.T., Scholtz C., Dilatancy in the fracture of crystalline rocks. *Journal of Geophysical Research*, *71*, 3939–3953 (1966).

Brown, D., Duchane, D., Heiken, G., Hriscu, V., *Mining the Earth's Heat: Hot Dry Rock Geothermal Energy*, Springer-Verlag, Berlin, Heidelberg (2012).

Cueto-Felgueroso, L., Santillán, D., Mosquera, J. C., Stick-slip dynamics of flow-induced seismicity on rate and state faults. *Geophysical Research Letters*, *44*(9), 4098–4106 (2017).

De Simone, S., Carrera, J., Vilarrasa, V., Superposition approach to understand triggering mechanisms of post-injection induced seismicity. *Geothermics*, *70*, 85–97 (2017).

Deichmann, N., Kraft, T., Evans, K. F., Identification of faults activated during the stimulation of the Basel geothermal project from cluster analysis and focal mechanisms of the larger magnitude events. *Geothermics*, *52*, 84–97 (2014).

Ellsworth, W. L., Giardini, D., Townend, J., Ge, S., Shimamoto, T., Triggering of the Pohang, Korea, earthquake (Mw 5.5) by enhanced geothermal system stimulation. *Seismological Research Letters*, *90*(5), 1844–1858 (2019).

Ghassemi, A., A review of some rock mechanics issues in geothermal reservoir development. *Geotechnical and Geological Engineering*, *30*(3), 647–664 (2012).

Giardini, D., Geothermal quake risks must be faced. *Nature*, *462*(7275), 848–849 (2009).

Häring, M. O., Schanz, U., Ladner, F., Dyer, B. C., Characterisation of the Basel 1 enhanced geothermal system. *Geothermics*, *37*(5), 469–495 (2008).

Hsieh, P. A., Deformation-induced changes in hydraulic head during ground-water withdrawal. *Groundwater*, *34*(6), 1082–1089 (1996).

Kamali, A., Ghassemi, A., Analysis of natural fracture shear slip and propagation in response to injection, *Proceedings of 41st Workshop on Geothermal Reservoir Engineering*, Stanford, CA (2016).

Kim, K., Makhnenko, R. Y., Coupling between poromechanical behavior and fluid flow in tight rock. *Transport in Porous Media*, *135*(2), 487–512 (2020).

Kim, K. H., Ree, J. H., Kim, Y., Kim, S., Kang, S. Y., Seo, W., Assessing whether the 2017 M w 5.4 Pohang earthquake in South Korea was an induced event. *Science*, *360*(6392), 1007–1009 (2018).

Lepillier, B., Yoshioka, K., Parisio, F., Bakker, R., Bruhn, D. Variational phase-field modeling of hydraulic fracture interaction with natural fractures and application to enhanced geothermal systems. *Journal of Geophysical Research: Solid Earth*, *125*(7), e2020JB019856 (2020).

Majer, E. L., Baria, R., Stark, M., Oates, S., Bommer, J., Smith, B., Asanuma, H., Induced seismicity associated with enhanced geothermal systems. *Geothermics*, *36*(3), 185–222 (2007).

McClure, M.W., Horne, R.N., An investigation of stimulation mechanisms in enhanced geothermal systems. *International Journal of Rock Mechanics and Mining Sciences*, *72*, 242–260 (2014).

McGarr, A., Maximum magnitude earthquakes induced by fluid injection. *Journal of Geophysical Research*, *119*, 1008–1019 (2014).

Mukuhira, Y., Asanuma, H., Niitsuma, H., Häring, M. O., Characteristics of large-magnitude microseismic events recorded during and after stimulation of a geothermal reservoir at Basel, Switzerland. *Geothermics*, *45*, 1–17 (2013).

Mukuhira, Y., Dinske, C., Asanuma, H., Ito, T., Häring, M. O., Pore pressure behavior at the shut-in phase and causality of large induced seismicity at Basel, Switzerland. *Journal of Geophysical Research: Solid Earth*, *122*(1), 411–435 (2017).

Niemz, P., Cesca, S., Heimann, S., Grigoli, F., von Specht, S., Hammer, C., Zang, A., Dahm, T., Full-waveform-based characterization of acoustic emission activity in a mine-scale experiment: A comparison of conventional and advanced hydraulic fracturing schemes. *Geophysical Journal International*, *222*(1), 189–206 (2020).

Olivella, S., Alonso, E. E., Gas flow through clay barriers. *Géotechnique*, *58*(3), 157–176 (2008).

Olivella, S., Carrera, J., Gens, A., Alonso, E. E., Non-isothermal multiphase flow of brine and gas through saline media. *Transport in Porous Media*, *15*, 271–293 (1994).

Olivella, S., Gens, A., Carrera, J., Alonso, E. E., Numerical formulation for a simulator (CODE_BRIGHT) for the coupled analysis of saline media. *Engineering Computations*, *13*, 87–112 (1996).

Pampillon, P., Santillán, D., Mosquera, J. C., Cueto-Felgueroso, L., Dynamic and quasi-dynamic modeling of injection-induced earthquakes in poroelastic media. *Journal of Geophysical Research: Solid Earth*, *123*(7), 5730–5759 (2018).

Parisio, F., Yoshioka, K., Modeling fluid reinjection into an enhanced geothermal system. *Geophysical Research Letters*, *47*(19), e2020GL089886 (2020).

Rubin, A. M., Ampuero, J. P., Earthquake nucleation on (aging) rate and state faults. *Journal of Geophysical Research: Solid Earth*, *110*(B11), 1–24 (2005).

Segall, P., Lu, S., Injection-induced seismicity: Poroelastic and earthquake nucleation effects. *Journal of Geophysical Research: Solid Earth*, *120*(7), 5082–5103 (2015).

Verdon, J. P., Bommer, J. J., Green, yellow, red, or out of the blue? An assessment of Traffic Light Schemes to mitigate the impact of hydraulic fracturing-induced seismicity. *Journal of Seismology*, *25*(1), 301–326 (2021).

Vilarrasa, V., De Simone, S., Carrera, J., Villaseñor, A., Unraveling the causes of the seismicity induced by underground gas storage at Castor, Spain. *Geophysical Research Letters*, *48*(7), e2020GL092038 (2021).

Vilarrasa, V., Koyama, T., Neretnieks, I., Jing, L. Shear-induced flow channels in a single rock fracture and their effect on solute transport. *Transport in Porous Media*, *87*(2), 503–523 (2011).

Xie, L., Ki-Bok, M., Yoonho, S., Observations of hydraulic stimulations in seven enhanced geothermal system projects. *Renewable Energy*, *79*, 56–65 (2015).

Xie, L., Min, K. B., Discrete element modelling of multi-stage hydroshearing stimulation in enhanced geothermal system. *Geothermics*, *95*, 102135 (2021).

Yeo, I. W., Brown, M. R. M., Ge, S., Lee, K. K., Causal mechanism of injection-induced earthquakes through the Mw 5.5 Pohang earthquake case study. *Nature Communications*, *11*(1), 2614 (2020).

Yeo, I. W., De Freitas, M. H., Zimmerman, R. W., Effect of shear displacement on the aperture and permeability of a rock fracture. *International Journal of Rock Mechanics and Mining Sciences*, *35*(8), 1051–1070 (1998).

Zang, A., Zimmermann, G., Hofmann, H., Stephansson, O., Min, K. B., Kim, K. Y., How to reduce fluid-injection-induced seismicity. *Rock Mechanics and Rock Engineering*, *52*(2), 475–493 (2019).

Zareidarmiyan, A., Parisio, F., Makhnenko, R. Y., Salarirad, H., Vilarrasa, V., How equivalent are equivalent porous media? *Geophysical Research Letters*, *48*(9), e2020GL089163 (2021).

Chapter 8

Enhanced drilling and recovery methods in geothermal reservoirs

Berker Polat, Shahin Jamali, and Volker Wittig
Fraunhofer IEG

CONTENTS

8.1 GEOTHERMAL RESERVOIRS AND EGS

Enhanced (or engineered) geothermal systems mostly refer to petrothermal reservoirs, targeting the exploitation of heat from dry and insufficient or little natural permeability formations, i.e., so-called hot dry rock. This play type or source represents the vast majority of geothermal resources in the subsurface around the world. Essentially, EGS-type geothermal systems portray a large-scale downhole heat exchanger, functioning by injecting fluid into the subsurface under carefully engineered and controlled conditions into a by then artificially created reservoir, meaning necessary rock permeability has previously been altered and enhanced for the cycling of the work fluid through the reservoir. This substantially increased fluid-formation contact, comparable to downhole borehole heat exchangers, renders this type of geothermal system extremely capable. Moreover, its universal availability and virtual inexhaustibility deem it more viable than other forms of geothermal energy for the purpose of district heating and electricity generation for baseload energy demand (Stober and Bucher, 2014; Tester et al., 2006).

Two main entities comprise an underground geothermal reservoir system: an artificial wellbore and a natural reservoir. A good connection between the two is paramount for a viable, sustainable production or injection system, rendering the interface of the wellbore and reservoir a crucial part of the overall subsurface system. However, there are occurrences that impede this connection, a major one being mud infiltration during the drilling process or naturally low permeability within reservoirs. And there are certain processes and treatments that have great potential to "repair" and enhance this connection, such as wellbore perforation and reservoir stimulation through hydraulic, chemical, mechanical, or thermal means. To facilitate this connection, a

DOI: 10.1201/9781003271475-8

workover process for such reservoir stimulation known as radial jet drilling (RJD) has been under development and is increasing in application in hydrocarbon as well as recently in geothermal systems, helping to create such engineered geothermal systems and make hot dry rock reservoirs more viable and productive. Figure 8.1 shows hydraulic, matrix, and RJD stimulation types.

8.2 RADIAL JET DRILLING TECHNOLOGY

Radial jet drilling (RJD) is a well-proven reservoir stimulation technology that targets enhancing the immediate wellbore-reservoir contact by eroding numerous, mainly radially oriented laterals into the formation and thus creating micro-sized drainholes to enlarge the production (or injection) section. By doing so, the vicinity of the wellbore is penetrated well beyond, e.g., by up to 100 meters into the reservoir, thereby increasing productivity or injectivity out of the main wellbore and its surrounding reservoir. RJD technology encompasses the following multitude of technologies: directional drilling, downhole bit drive, hydraulic rock erosion and milling, ultra-short radius wellbore deviation, wellbore sidetracking, coiled tubing, and jointed pipe. The detailed process of this technique is briefly explained below (Blöcher et al., 2016).

In a beginning step, a deviation tool similar to a conventional whipstock, though much smaller, is lowered into an existing borehole, whether cased or open, by means of jointed pipe requiring a completion or workover rig. In a second step, coiled tubing is run into the hole to deploy a downhole bit drive and a high-pressure hydraulic rock erosion tool through the deviation tool. Finally, the deviation tool is used to sidetrack from the borehole into the reservoir with the downhole hydraulic rock erosion tool. This likewise commences the stimulation treatment and the creation of the micro drainholes mentioned above (Ahmed, 2017; Blöcher et al., 2016; Bruni et al., 2007; Buset et al., 2001; Dickinson et al., 1989; Dickinson and Dickinson, 1985; Ragab and Kamel, 2013).

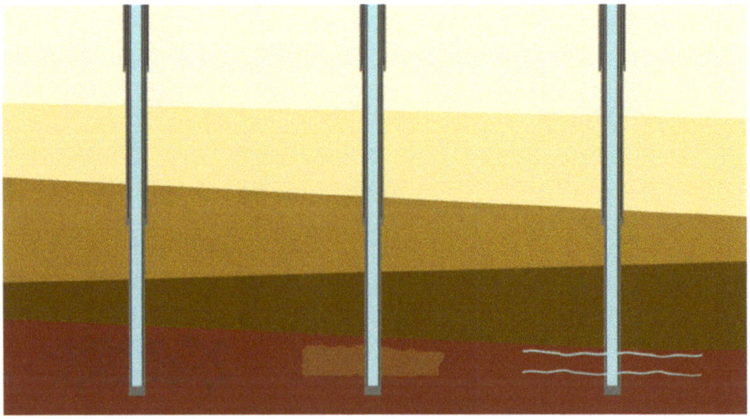

Figure 8.1 EGS enhancement scenarios within geothermal reservoirs: hydraulic leak off or fracturing (left), chemical matrix treatment (middle), and mechanical RJD (right) (Polat, 2020).

8.3 RADIAL JET DRILLING: PROCESS

Throughout the workover with the RJD process, the deviation tool is suspended and held in place horizontally by the jointed pipe. As part of the bottom hole assembly, the deviation tool is usually placed in-between the centralizing and stabilizing downhole elements, thus providing a smooth and good interface between the deviation tool and the borehole wall. The downhole bit drive and hydraulic rock erosion tool are powered by the water or working fluid used for the jetting operation. The bit drive thereby is a downhole motor providing torque and RPM to a milling bit on a flex shaft for cutting and milling the required hole in the casing (also see below). After casing milling, the BHA must be changed to a high-pressure nozzle system, featuring forward- and backward-oriented orifices or nozzles where the working fluid will exit. The forward or front jets erode the rock mass, whereas the fluid exiting the latter, backward-oriented "thruster" jets propels and advances the jetting BHA further into the rock formation or reservoir. Therefore, it is a self-propelling system, not requiring any active WOB from the tool string or coil tubing above. The nozzle, or jetting BHA, is attached to a rather small and very flexible hose, which in turn is attached to the coiled tubing above, supplying the working fluid from the surface. The jetting BHA with its hose is deviated to initiate the sidetrack, with the latter trailing the former and both being in tension throughout the entire jetting process, creating the new lateral micro drainhole. Weight on bit does not need to be applied as mentioned, as the nozzle BHA produces its own thrust and may not physically touch the rock surface when trying to conduct the fastest and most efficient hydraulic rock erosion process. At all times, the fluid exiting the backward thruster nozzles must produce a greater force than the reactive force evoked by the fluid exiting the forward-oriented, rock-eroding orifices and the increasing friction force of the hose being pulled behind. The penetration and sidetrack creation halt and come to an end as both forces reach equilibrium. Figure 8.2 shows the RJD process with its main components (Ahmed, 2017; Bin et al., 2016; Blöcher et al., 2016; Bruni et al., 2007; Buset et al., 2001; Ragab and Kamel, 2013; Saikia et al., 2019).

Figure 8.2 Main downhole components of a Radial Jet Drilling BHA.

In cased hole applications, the casing or liner must be penetrated prior to the following hydraulic rock erosion of the formation. This is conducted using a downhole bit drive powered by a conventional small-diameter positive displacement motor, a flex shaft, and a milling bit, where the former powers the latter two with rotation. The flex shaft and bit are deviated to mill a window through casing and cement, while the positive displacement motor remains in the borehole section. For application in a cased-hole wellbore, RJD requires respective trips of the coiled tubing for casing milling and drainhole creation. Thereby, drainhole creation resumes where casing milling and window initiation previously came to a halt (Blöcher et al., 2016; Bruni et al., 2007; Ragab and Kamel, 2013; Saikia et al., 2019).

Moreover, the jetting operation targets the creation of numerous microdrainholes under various orientation angles at any one or various depth levels. Four to six drainholes are commonly created on any one horizon, which is made feasible by changing the deviation tool orientation. Subsequently, the pipe string is rotated from the surface, resulting in a change in the tool orientation. This is conducted after the retrieval of the coiled tubing and completion of each micro drainhole. If required, a gyroscope can provide further information about the exact downhole orientation in a separate run. In such cases, the gyroscope is lowered through the pipe to the deviation tool by means of wireline, facilitating real-time surface readout. Reorientation of the deviation tool is not possible without this information. However, doing so will result in a lack of information regarding the new exact position or orientation of the tool, as the surface rotation of the pipe does not necessarily correspond to its downhole rotation due to torsional spring loading effects.

Furthermore, this creation of drainholes may be repeated at several other depth levels to successfully and increasingly connect the borehole to the reservoir. For this, upon completion of one level, the deviation tool is pulled up or lowered further down to reach a new desired horizon to be completed with micro sidetracks (Blöcher et al., 2016; Bruni et al., 2007; Saikia et al., 2019).

8.4 DEVELOPMENTS OF RADIAL JET DRILLING

Various parameters need to be considered, which greatly affect the efficiency of any RJD process. The viability of possible rock erosion via jetting is severely and primarily influenced by nozzle design and, secondarily, operational parameters such as fluid flow rates, resulting in differential pressure and jet speed. As shown in Figure 8.3, nozzles may be of rotating or static type, producing differently shaped drainholes or notches, respectively, as shown in Figure 8.4. The fluid's exit velocity and especially the distance between the nozzle orifice or outlet and the rock surface (standoff distance) also significantly affect and determine the outcome of any rock erosion process. Moreover, phenomena like cavitation and pulsation within the fluid enhance and influence the erosion process even more. The fluid composition itself does not have a significant influence, as RJD is usually conducted using clear water unless abrasives are being used and added to the fluid (Blöcher et al., 2016; Farmer and Attewell, 1965; Maurer and Heilhecker, 1969).

Figure 8.3 Examples of investigated rotating and static nozzles at IEG.

At Fraunhofer IEG in Bochum, Germany, adaptation and further development of RJD to overcome the challenges set by the requirements within geothermal reservoirs have been underway for many years. Research in this area focuses especially on the applicability of RJD in hard rock formations and the higher temperatures common to geothermal prospects like hot dry rock. By doing so, system pressure levels have to be increased, and the jetting process is holistically optimized to foster improved and enhanced rock penetration. This includes keeping pressure losses in the system at a minimum, novel nozzle design for rock erosion, e.g., adding pulsation, cavitation, and even mechanical milling support, and downscaling of operational requirements for economic reasons. Therefore, novel approaches for the creation of microdrainholes have been investigated.

These developments are not limited merely to the hydraulic rock erosion process; they also make use of mechanically supported downhole jetting or milling bits of various kinds and even look into small-size, compact rotary-type percussion drilling techniques. The goal also remains to optimize surface-to-downhole transfer of hydraulic energy through the coiled tubing system and thus provide for direct or indirect (with mechanical support) utilization for an efficient rock disintegration process. In sedimentary-type rocks, direct hydraulic rock erosion typically results in high rates of penetration, whereas indirect (turbine milling or percussion) mechanically supported jetting milling or drilling processes are more useful and effective in hard rocks.

Furthermore, the number of coiled tubing round trips ("tripping") must be reduced and kept to a minimum, as each tripping operation is timely and increases non-productive time and the cost of operations. This is often still a bottleneck in downhole geothermal applications. Ultimately, research work at IEG, novel methods, and further optimization

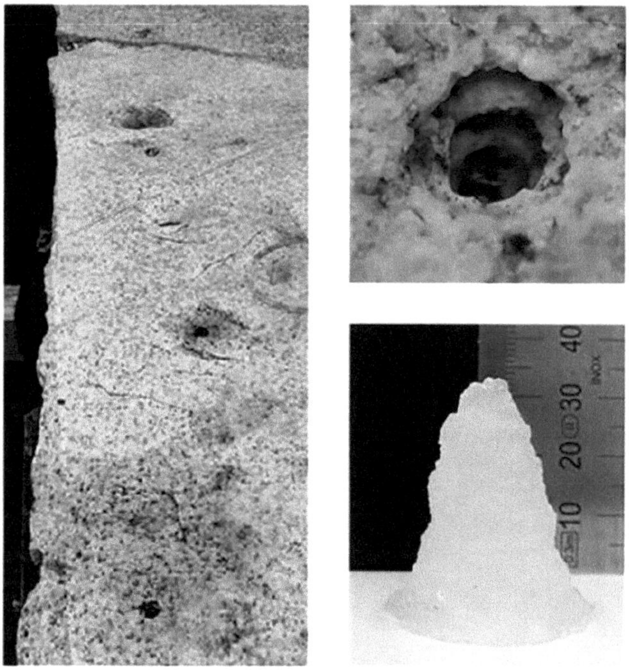

Figure 8.4 RJD testing in Grimsel granite at IEG: produced cavities and notches analyzed in the lab.

Figure 8.5 RJD notch eroded into the side of an open hole main wellbore wall (Grimsel Granite).

will facilitate the easy and timely completion of several drainholes on any one level or depth through one coiled tubing trip only. Research and development is currently ongoing, including making RJD wireline compatible and many laboratory experiments simulating downhole reservoir conditions as well as field tests under atmospheric conditions and full-scale field demonstrations within industry applications.

So far, laboratory experiments have allowed the alteration and investigation of parameters like flow rate, standoff distance, and pulsation, resulting in favorable outcomes under simulated reservoir-type conditions down to 2,000m depth. Under atmospheric conditions, pressure regimes up to 1.000 bars have yielded drainholes and good excavation volumes in very hard rocks like granites. Furthermore, field demonstrations proved the applicability of the technology in larger geothermal settings. Figure 8.5 shows an eroded notch in granite in the Bedretto Underground Laboratory BUL of the Gotthard mountain range, Switzerland.

8.5 RADIAL JET DRILLING: MONITORING

Any RJD operation so far is a rather inexpensive and very environmentally friendly alternative to other stimulation treatments, although it is still not advanced enough to make good use of its full potential and effectiveness. One step lies in the assessment of the drainholes or cavities produced upon completion of such a stimulation treatment. Such lateral holes can actually be accessed through surface readout by innovative monitoring or post-run by logging and camera runs, all of which are being extensively investigated and used at Fraunhofer IEG. In-situ logging and monitoring are based on acoustic emission, including machine learning and AI-based real-time analysis, whereas post-logging utilizes conventional borehole logging techniques fitted to the radial jetting microscale.

Acoustic emission analysis is a non-destructive, passive monitoring method encompassing the detection of elastic waves generated during processes and their conversion to electrical signals for further processing. The emissions are a result of rapid releases of localized stress energy within materials such as rocks, rendering acoustic emission analysis a non-destructive monitoring method as no external energy is applied to the material. The emissions are measured in various ways, namely by utilizing piezoelectric transducers, which are mainly a result of micro-cracks that form within the material. In formation rocks, these cracks are evoked by thermo-, chemo-, or hydro-mechanical loads. Acoustic emission analysis conventionally focuses mainly on event count and dominant frequency characterization; however, novel approaches such as pattern recognition and machine learning were employed to adapt this technology to better understand the rock erosion process found in RJD. This is because a single acoustic emission wave may present valuable information outlining an action or process. The idea is to single out events for different processes and various rock types, as acoustic emissions are unique. This approach allows the identification of rock type, rock erosion and failure, and many more, resulting in a proper monitoring method and a better understanding of the RJD process. In combination with surface readouts, acoustic emission monitoring can facilitate active operational decision-taking and process modifications in an otherwise passive RJD operation. Many laboratory experiments with various rock types and parameter alterations have been conducted to study their respective acoustic emission responses and feed the local database (Grosse and Ohtsu, 2008; Lei et al., 2000; Liptai et al., 1970; Matthews, 1983; Mba, 2006; Ohnaka and Mogi, 1982; Sikorska and Mba, 2008; Tripathi et al., 2016).

In-situ, on-the-go logging of the RJD process is vital for success, high efficiency, and assessment with possible control of the jetting operation and its successful outcome. Furthermore, drainhole homogeneity, depth of penetration, hole caliper, and pathway or trajectory of a generated lateral provide vital information about the overall success of the operation and allow for good comparison to other stimulation technologies, rock types, and reservoirs. However, any (downhole) sensor, monitoring, and control effort also adds to the complexity and, thus, cost of such an otherwise very cheap and simple jetting operation. In most cases, the added cost of such control equipment will far outweigh the gain in success and improved production of geothermal wells.

Design criteria and considerations for such logging tools have to address their utilization in a very compact environment, deviated or near horizontal orientation, high temperatures, and memory-based or surface readout options. However, the added operational complexities of such tools may lie in the necessity to either add real-time, in-situ data transfer during jetting or, for the post-run option, have additional coiled tubing or wireline trips to monitor any cavities or laterals created. However, naturally, it is nearly impossible to re-enter laterals or drainholes once the deviation tool has been reoriented or moved, rendering a single logging trip for all drainholes at any one level or depth impossible. Thus, possible "quality control" type post-run logging trips must be conducted subsequent to each jetting step, possibly adding many round trips and with that non-productive time to the overall RJD process. Though monitoring and control activities add to the overall jetting operation, as mentioned above, they also deliver and provide a higher-quality jetting operation as an outcome in the end.

REFERENCES

Ahmed, H.K. (2017) A technical review of radial jet drilling. *Journal of Petroleum and Gas Engineering*, [Online] 8 (8), 79–89. Available from doi:10.5897/JPGE2017.0275.

Bin, W., Gensheng, L., Zhongwei, H., Jingbin, L., Dongbo, Z. & Hao, L. (2016) Hydraulics calculations and field application of radial jet drilling. *SPE Drilling & Completion*, 31 (01), 71–81. doi:10.2118/179729-PA

Blöcher, G., Peters, E., Reinsch, T., Petrauskas, S., Valickas, R. & van den Berg, S. (2016) *D3.2 Report on Radial Jet-Drilling (RJD) Stimulation Technology*. GFZ German Research Centre for Geosciences, Potsdam.

Bruni, M.A., Biasotti, J.H. & Salomone, G.D. (eds.) (2007) *Radial Drilling in Argentina*. SPE, Buenos Aires.

Buset, P., Riiber, M. & Eek, A. (eds.) (2001) *Jet Drilling Tool: Cost-Effective Lateral Drilling Technology for Enhanced Oil Recovery*. SPE, Houston, TX.

Dickinson, W., Anderson, R.R. & Dickinson, R.W. (1989) The ultrashort-radius radial system. *SPE Drilling Engineering*, 4 (03), 247–254.

Dickinson, W. & Dickinson, R.W. (eds.) (1985) *Horizontal Radial Drilling System*. SPE, Bakersfield, CA.

Farmer, I.W. & Attewell, P.B. (1965) Rock penetration by high velocity water jet. *International Journal of Rock Mechanics and Mining Sciences & Geomechanics Abstracts*, 2 (2), 135–153.

Grosse, C. & Ohtsu, M. (2008) *Acoustic Emission Testing*. Berlin, Heidelberg: Springer Berlin Heidelberg.

Lei, X., Kusunose, K., Rao, M.V.M.S., Nishizawa, O. & Satoh, T. (2000) Quasi-static fault growth and cracking in homogeneous brittle rock under triaxial compression using acoustic emission monitoring. *Journal of Geophysical Research: Solid Earth*, [Online] 105 (B3), 6127–6139. Available from doi:10.1029/1999JB900385.

Liptai, R.G., Harris, D.O., Engle, R.B. & Tatro, C.A. (1970) *Acoustic Emission Techniques in Materials Research*. California Univ., Livermore Lawrence Radiation Lab, San Antonio, TX.

Matthews, J.R. (ed.) (1983) *Acoustic Emission*. New York, NY: Gordon and Breach.

Maurer, W.C. & Heilhecker, J.K. (eds.) (1969) *Hydraulic Jet Drilling*. SPE, Austin, TX.

Mba, D. (2006) Development of acoustic emission technology for condition monitoring and diagnosis of rotating machines: Bearings, pumps, gearboxes, engines, and rotating structures. *The Shock and Vibration Digest*, [Online] 38 (1), 3–16. Available from doi:10.1177/0583102405059054.

Ohnaka, M. & Mogi, K. (1982) Frequency characteristics of acoustic emission in rocks under uniaxial compression and its relation to the fracturing process to failure. *Journal of Geophysical Research: Solid Earth*, [Online] 87 (B5), 3873–3884. Available from doi:10.1029/JB087iB05p03873.

Polat, B. (2020) *Investigation of Deviation Mechanisms for Micro Drilling and Stimulation Tools*. Clausthal-Zellerfeld. Technical University Clausthal, TUC 2020.

Ragab, A.M. & Kamel, A.M. (eds.) (2013) *Radial Drilling Technique for Improving Well Productivity in Petrobel-Egypt*. SPE, Cairo.

Saikia, P.P., Dutta, U.A., Tumung, R., Verma, S., Ahmed, A.U. & Mukerjee, A. (eds.) (April 2019) *Application of Radial Jet Drilling Technique with Gyro System to Enhance Reservoir Potential in Mature Sand Stone Reservoir in Assam-Arakan Basin Candidate Selection to Execution: A Detailed Case Study*. SPE, Mumbai.

Sikorska, J.Z. & Mba, D. (2008) Challenges and obstacles in the application of acoustic emission to process machinery. *Proceedings of the Institution of Mechanical Engineers, Part E: Journal of Process Mechanical Engineering*, [Online] 222 (1), 1–19. Available from doi:10.1243/09544089JPME111.

Stober, I. & Bucher, K. (2014) *Geothermie*. 2nd edn. Berlin, Heidelberg: Springer Berlin Heidelberg.

Tester, J.W., Anderson, B.J., Batchelor, A.S., Blackwell, D.D., DiPippo, R., Drake, E.M., John, G. & Livesay, B. (2006) *The Future of Geothermal Energy: Impact of Enhanced Geothermal Systems (EGS) on the United States in the 21st Century an Assessment.* Idaho Falls, ID: National Laboratory.

Tripathi, R., Srivastava, M., Hloch, S., Adamčík, P., Chattopadhyaya, S. & Das, A.K. (2016) Monitoring of acoustic emission during the disintegration of rock. *Procedia Engineering,* [Online] 149, 481–488. Available from doi:10.1016/j.proeng.2016.06.695.

Chapter 9

International cooperation to address and mitigate the climate change issue using unconventional geothermal technology (EGS)

Roy Baria
Mil-Tech UK Ltd

Joerg Baumgaertner
BESTEC GmbH

Hylke Glass
University of Exeter

Andrew Jupe
Altcom Ltd

Ann Robertson-Tait
GeothermEx

Graeme Beadsmore
Hot Dry Rocks Pty

Corinna Abesser
BGS

Dimitra Teza
BESTEC GmbH

Hiro Asanuma
FREA

Igor Kocis
GA Drilling

DOI: 10.1201/9781003271475-9

CONTENTS

9.1 INTRODUCTION

Changes in the world's climate are well documented and scientifically supported in many reputable publications. The consequences of this for many countries and communities are apparent in the spurious floods, fires, drought, more frequent and intense natural disasters, and the rise in sea level, which may decimate many island-based communities. Unfortunately, it is the poorest and most vulnerable who are and will be affected the most.

The World Bank has confirmed that the risks of unabated climate change are mounting with each passing year; unless urgent action is taken, the impact of climate change could push an additional 100 million people into poverty by 2030 (https://www.worldbank.org/en/news/feature/2015/11/08/rapid-climate-informed-development-needed-to-keep-climate-change-from-pushing-more-than-100-million-people-into-poverty-by-2030). Future temperature projections indicate that more than 140 million people who live near the equator will experience temperatures in excess of 50°C by 2050. These people will be forced to migrate north or south, leading to the distinct possibility of conflicts between nations and the destabilisation of these regions.

The effect of climate change is already apparent, has a measurable impact on human health and is expected to grow. Air pollution is expected to be responsible for 7 million premature deaths per year, and the increased cost of health care is estimated to be as much as $4 billion per year. Additionally, the impact of extreme natural disasters is estimated to cost around $520 billion in annual consumption and is likely to force some 26 million people into poverty each year. The World Bank estimates that the financing required for an orderly transition to a low carbon, energy-resilient global economy must be counted in the trillions of dollars (http://www.worldbank.org/en/topic/climatechange/overview).

Measures must be applied to either slow down the effect of climate change or reverse it if possible. Therefore, the use of heat and power generation from renewable sources is a constructive way forward.

Geothermal is a vast renewable resource whose exploitation has been limited to date. Currently, the predominant method of using the Earth's heat is the development of conventional hydrothermal systems using existing drilling and power generation technologies. Hydrothermal resources that have been developed and exploited are generally confined to areas characterised by volcanism and/or crustal extension and thinning; both lead to high heat flow and therefore the ability to access high-temperature resources with relatively shallow drilling. Economically, this is a distinct advantage because drilling costs can be relatively low and the resource can be relatively well defined. On the negative side, conventional hydrothermal resources are geographically limited, and population density in some of these specific areas is, on average, extremely low. In comparison to the total available geothermal energy resource base in the world, the current level of development of hydrothermal resources does not significantly contribute to mitigating climate change issues.

In a conventional hydrothermal system, meteoric water infiltrates from depth and migrates towards the surface via faults and other permeable pathways. In the process, the water is heated and may contain dissolved minerals, depending on the local geology. One or more permeable zones may then host a hot fluid reservoir, from which some fluid continues to rise towards the surface owing to the buoyancy of hot water. Traditionally, hydrothermal systems are recognised by their geologic settings, surface expressions (e.g., hot springs, geysers, fumaroles and altered ground) and/or shallow subsurface expressions (e.g., warm waters in shallow wells) and can be further identified and characterised by geophysical and geochemical surveys, typically followed by shallow and deep drilling. Wells are drilled into the permeable zones of the hydrothermal systems to extract the hot fluid. Once the heat has been extracted in the power plant process, the cooled fluid is reinjected into permeable zones at a suitable distance from the production area in the expectation that this fluid will then be reheated as it returns towards the production area, thereby recharging the hydrothermal system. As discussed above, geologic conditions suitable for the formation of hydrothermal resources have a limited geographic extent (mainly around the Ring of Fire, with some important exceptions).

In contrast, the geographic extent of areas suitable for EGS development is vast. There is a significant proportion of the upper crust that is hot but not sufficiently permeable to enable the development of conventional hydrothermal resources. In many parts of the world, intrusive rock (such as granite) occurs at depths where its temperature is suitable for the production of heat and power, enhanced by its good heat conduction properties and radiogenic composition. The concept of EGS was developed with the understanding that there is a significant proportion of the upper crust that is accessible by drilling and is hot but not sufficiently permeable to create a natural hydrothermal system. The EGS approach allows permeability to be enhanced, ideally by exploiting natural fault systems that have some permeability but not enough to allow geothermal fluids to be produced at commercial rates. Permeability enhancement increases reservoir storage and allows heat to be mined via injection and production.

It is proposed that climate change may be mitigated by international cooperation on the application of EGS technology worldwide (Baria et al., 2016), thus reducing the dependence on hydrocarbons and associated pollutants. It is proposed that a cooperation structure be established under a consortium consisting of bodies such as the International Energy Agency, the World Bank, the United Nations and others, providing the priority and credibility needed. Proposed steps are mapping of deep faults, characterising these faults using geophysical techniques to delineating the presence of suitable quantities of fluid at depth, improve drilling technology to assist economics and establish Centre of Excellence to share technology and to train scientists and engineers required to maintain the growth of this process.

9.2 WORLD ENERGY DEMAND AND SUPPLY SCENARIO

The largest source of energy for the Earth is the sun; approximately 220 W are available on each square metre of the surface of our planet when the sun shines. The global potential for solar power has been estimated by the IEA (https://www.iea.org/geco/data/) to be ~570 TWe in 2018.

As discussed below, the global geothermal potential is of a similar order of magnitude.

9.2.1 World supply of geothermal energy

A hydrothermal resource requires heat, fluid and permeability. Geothermal hot springs have been exploited for domestic use since ancient times, but geothermal exploitation for industrial use started in the early 19th century, when hydrothermal energy was used for extracting boric acid in Larderello (Italy). The first geothermal district heating system began operating in Boise, Idaho (USA), at the end of the 19th century, followed by Iceland in the 1920s. The first successful attempt to produce electricity from geothermal heat was achieved at the start of the 20th century in Larderello, and the first large-scale geothermal power plant was installed at Wairakei (New Zealand) in the late 1950s, followed by the first two units at The Geysers, California (USA), in 1962. Since then, installed geothermal electricity has increased steadily. The global geothermal power capacity in 2009 was 10.7 GWe and generated approximately 67.2 TWhe/yr of electricity at an average efficiency rate of 6.3 GWh/MWe (Bertani, 2010). A remarkable growth rate in geothermal energy occurred from 1980 to 1985, which was largely driven by the interest of the hydrocarbon industry and the initiation of favourable power pricing under contracts in the USA and elsewhere.

Almost all the electricity generated by geothermal energy is derived from hydrothermal power plants located predominantly in volcanic domains near the boundaries of tectonic plates (Figure 9.1) and in areas of crustal thinning; both regions are characterised by high temperatures at economically drillable depths. These systems are often characterised by surface manifestations that provide clues to the underlying resource, but not always.

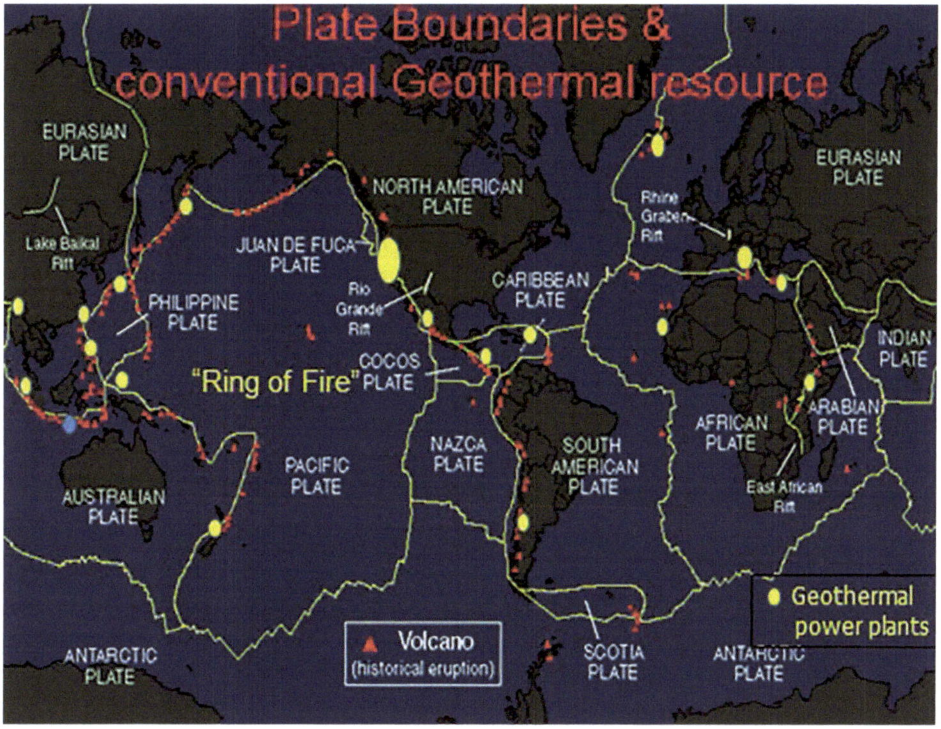

Figure 9.1 Plate boundaries, volcanoes and hydrothermal resources supplying electric power. Modified from the US Geothermal Education Office.

9.2.2 World energy demand

The intensity of energy demand in the world can be attributed to population density, the region's stage of industrialisation and the growth of the birth rate, all of which may fluctuate from decade to decade due to economic cycles and political changes. Although this can be a complex model to represent, a relatively simple indication would be to look at the population density map of the world to give some indication of areas of particularly high energy demand. The total world population is estimated to be around ~7.5 billion (UN estimate); a map showing population density is presented in Figure 9.2. As can be seen, the greatest demand for energy is associated with countries that have a high population density, such as China, Japan, India, Western Europe, the USA, Nigeria and Brazil.

9.2.3 The role of geothermal in energy supply and demand

The global potential for geothermal electricity has been estimated by Krewitt et al. (2009) at 45 EJ/yr - 12,500 TWhe, which is almost half of the 2018 global electricity generation based on data from the IEA. The Krewitt study also estimates that resources suitable for direct use can potentially supply 1,040 EJ/yr - 289,000 TWht and notes

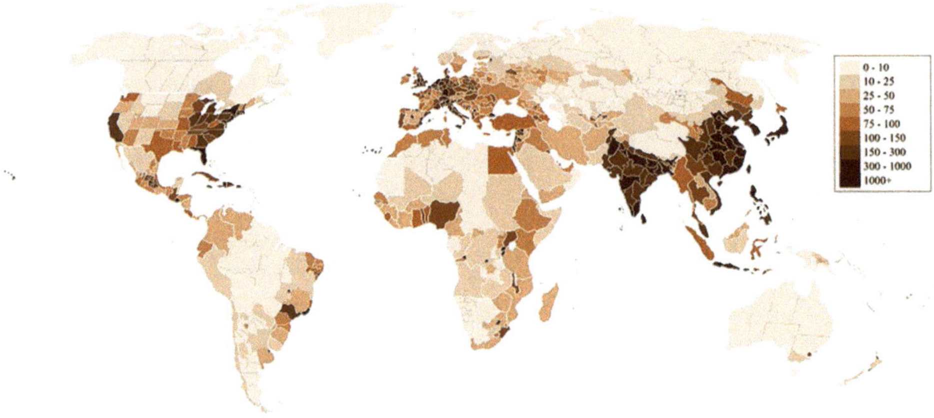

Figure 9.2 Global population density distribution (UN data base) shows that the demand for power in most areas of the world is not well aligned with areas that host conventional hydrothermal resources.

that worldwide final energy use for heat in 2008 was 159.8 EJ/44,392 TWht. The geothermal estimates presented by Krewitt et al. are conservative in that they exclude advanced geothermal technologies that could exploit hot rock or off-shore hydrothermal, magma and geopressured resources. Although geothermal energy has great technical potential, its exploitation is hampered by power price limitations (geothermal may or may not be the least-cost renewable source) and the limited geographic distribution of geothermal resources relative to centres of high energy demand (International Energy Agency, 2011)

Overlaying the map of current geothermal energy from hydrothermal resources (Figure 9.1) and the population density map (Figure 9.2) shows the disparity between areas of potential supply and areas of high demand. For example, New Zealand and Iceland can contribute an appreciable amount of hydrothermal energy, but both island nations have low populations (approximately 4.5 million and 350,000, respectively). Unlike commodities such as oil, gas, coal and timber, an important limitation of geothermal energy is that hot water cannot be conveyed over long distances or across continents; therefore, the limited availability of the hydrothermal resource does not lend itself to greater exploitation worldwide.

There is an abundant non-hydrothermal (unconventional) geothermal resource in the world; however, current methodologies for extracting this form of energy are limited, and costs are higher than for conventional hydrothermal resources. The idea of creating subsurface heat exchangers with large effective volumes by using hydraulic stimulation remains a valid technique. However, it has not yet been possible to create artificial heat exchangers that are sufficiently large to be sustainable for 30 or 40 years with acceptable flow rates, operating pressures (i.e., suitably low flow resistance) and far-field water losses in most EGS (closed system) projects to date. However, by realising these parameters in more common geologic settings than those hosting conventional hydrothermal resources, EGS can enable geothermal to become a global energy source.

9.3 THE ENGINEERED GEOTHERMAL SYSTEM (EGS)

9.3.1 Overview

The original concept of EGS (closed system), formerly known as hot dry rock (HDR), was developed in the early 1970s at Los Alamos in the USA. It consisted of drilling wells to a depth of 4,500m into the flank of a caldera at Fenton Hill, New Mexico, to access high temperatures (~300°C) and then to enhance the permeability of the system by injecting fluid under high pressure into fractures within the rock. This concept was replicated at the geothermal research project at Rosemanowes in the UK (Garnish, 1976; Parker, 1989; Willis-Richards et al., 1990) and at other sites around the world (Abe et al., 1999). The wells at the Rosemanowes site were drilled into granite to a depth of 2,000–2,500m, and the project was specifically designed to understand the physics of creating enhanced permeability in igneous rock (Pine et al., 1987), where shearing of natural fractures was the main mechanism for enhancing permeability.

Knowledge gained from more than 30 years of research in these early EGS projects formed the basis upon which the European project at Soultz was built (Abe et al., 1999; Baria, 1990; Baria and Green, 1990; Baria et al., 1989, 1999, b, 2000; Baumgärtner et al., 1995, 1998; Evans, 2005; Gérard et al., 1997; Valley, 2007). For the first time, wells were drilled successfully to 5,000m depth in a granite basement to access temperatures of 200°C. Although the Soultz project started as an HDR project (closed system EGS) with the necessity to carry out hydraulic stimulations to enhance permeability, the discovery of hydraulically conductive natural faults at greater depth and the relationship of these faults to the in-situ stresses opened up a new concept – open system EGS – for extracting energy from depth at higher temperatures. It became clear that natural, hydraulically conductive faults were more prevalent than had been originally anticipated. At Soultz and Rosemanowes, it was clear that natural fractures and faults were the dominant flow structures. In this way, they are similar to conventional hydrothermal systems; however, the permeabilities required for high flow rate wells were not present, and therefore, there remained a need to engineer the system to enhance permeability.

9.3.2 EGS potential in Cornwall

Cornwall has a history of deep mining in the UK, yielding a database that gives some indication of the fault systems encountered at depth and their characteristics. The Cornubian Batholith, which underlies much of Cornwall and west Devon, is the largest of the granite batholiths in the UK. It is exposed, from west to east, in six major plutons, namely the Isles of Scilly, Land's End, Tregonning-Godolphin, Carnmenellis, St Austell, Bodmin Moor and Dartmoor granites. All the major plutons have been dated to be Early Permian and were emplaced over a period of approximately 20 Ma, from ~294 Ma (Carnmenellis Granite), ~285 Ma (St Austell Granite) to ~274 Ma (Land's End Granite).

The hydrogeological (porosity and permeability) characteristics of a granite rock mass are dictated by its cooling history, tectonic history and depth (including times at which it may have been exposed at the ground surface) and thus may vary from highly permeable within near-surface weathering zones to significantly lower effective primary permeability at depth, confined within discontinuities and fractures.

The contemporary stress regime in Cornwall and west Devon is similar to that across much of northwest Europe, with the maximum horizontal stress (σH) oriented approximately NNW-SSE. These fracture zones are of particular importance for the development of EGS reservoirs in southwest England as they are near optimally oriented with respect to the in-situ stress regime, regional in extent and occur throughout the Devonian-Carboniferous successions and the granites.

Fracture formation and reactivation during granite magmatism and related mineralisation were primarily controlled by the evolving post-Variscan tectonic regime (Dine, 1956; Garnett, 1961; Shail and Alexander, 1997; Hawkes, 1981). These fractures facilitated the migration and mixing of magmatic, meteoric and basial fluids that were fundamental to the mineralisation process. The main period of hydrothermal mineralisation was during the late/post-granitic emplacement. The dominant trend for the mineralised lodes is ENE-WSW to E-W, and NNW-SSE for the mineralised faults. The mineralised fracture systems have been interpreted to result from interactions between regional stresses and magmatic fluid pressures, typically resulting in extensional faults (Hosking, 1949).

Crosscourse is the name given to the faults that occur in a general N-S direction. They can be classified into two main types:

a. **Fissure fill crosscourses** – extensional fissures filled with clay and quartz, having a "vuggy" nature (sometimes named "cross-veins"),
b. **Shear/wrench crosscourses** – zones of intense micro-shearing in which feldspars and micas have undergone advanced argillic alteration into a clay gouge infill ("fluccans").

Crosscourses are generally steeply inclined (70°–80°) and vary from individual veins that are less than 1 m in width with minimal offset of wall rock features to fault zones greater than 100 m in width that may offset tin-copper lodes by more than 100 m. Spacing for fault zones with trace lengths of more than 1 km can vary from between 1 and 10+ km, but between 10 and 500 m for major veins. Crosscourses existed as fault fissures prior to lode mineralisation and were an important component of structural control during mineralisation.

Crosscourse mineralisation is primarily related to the migration of high-salinity basinal fluids from formerly overlying and adjacent Permo-Triassic Basins (Scrivener et al., 1994; Mueller et al., 1999). Migration of highly saline brines was largely controlled by NW-SE faults, and there was only limited mixing with meteoric fluids. Crosscourse mineralisation occurs in several broad paragenetic associations, including (i) tin, lead, copper, zinc and fluorite; (ii) uranium, cobalt, silver and arsenic; and (iii) iron. The gangue (ore host rock) mineralogy typically comprises one or more of quartz, chalcedony, carbonates, fluorite, hematite and clay, often with banding and vuggy cavities. It is apparent that some of this mineralisation occurred during the Mid- to Late Triassic and is related to regional fluid flow events. The development and reactivation of the crosscourse fracture network reflect the onset of a major episode of Triassic rifting that affected much of Western Europe, where the dominant extension direction was ENE-WSW, which resulted in the development of NW-SE–oriented faults.

Water ingress proved to be a major problem in the Cornish mines, which necessitated expensive pumping as the mines deepened. Evidence from historic underground mining activity shows that the greatest proportion of water that enters shallow mine workings is derived from the surface; however, warm saline fluids have been observed in some of the deeper mines. It is the crosscourses, lodes and elvans (hard intrusive rocks found in Cornwall, typically quartz porphyry), which are commonly the high-permeability structures that enable the preferential flow of meteoric and saline fluids through the rock. In addition, rhyolite (elvan) dykes, which are commonly associated with lode structures, often contain microfractures that result in high permeability and groundwater flow in mine workings. Some of the main crosscourses that have been encountered in mines are associated with groundwater flow to significant depths. They commonly provide natural pathways for the passage of groundwater and are notorious as sources of inflow into mine workings and for hydraulic connectivity between mine workings. Fluid inflows typically occur in fractured zones that lie adjacent to and sub-parallel to the main crosscourses. Thermal springs in mines were commonly associated with crosscourses, for example, on the 820m level at South Crofty Mine.

Observations in deep mines in Cornwall (Figure 9.3) have shown that these faults at depth can produce significant flow (up to 700 L/s) for infinitum. The presence of these faults is apparent in the satellite images and surface expression (Figure 9.4). These faults are linked to the geomechanical properties of the rock mass and are aligned around the direction of maximum horizontal stress. A good knowledge of geomechanics and the stress field is therefore crucial to understanding the preferential direction of fluid flow and enhancing the properties of these large conductive faults.

9.3.3 The open EGS system

The presence of large conductive faults at depth demonstrates that rather than creating an artificial heat exchanger using hydraulic stimulation of an impermeable jointed rock mass (closed system), it is more advantageous to use existing natural hydraulically conductive faults to mimic a natural hydrothermal system (open system). Where correctly aligned in relation to the direction of maximum horizontal stress and with some hydraulic stimulation, these deep faults have the potential to behave like a hydrothermal system by permitting a reasonable sustained flow rate over a long period.

The configuration of injection and production wells is strongly influenced by the natural stress regime. In-situ conditions vary between localities, making it difficult to produce a pre-programmed model on which to design deep geothermal reservoirs. Therefore, until more data are available from projects in specific geologic regions, each system needs to be assumed to be site-specific and appropriate engineering manipulations are required to bring each system into successful production. As an example of the learning curve effect, the success of this technology at the Soultz project led to the development of the first commercial EGS plant within the northern Rhine Graben at Landau (Baumgärtner et al., 2013b) in Germany.

These pioneering R&D projects have led to a number of successful commercial EGS projects being established in Germany (Insheim, Landau) and France (ECOGI; Rittershoffen), all of which are located in the Rhine Graben. It is apparent now that deep faults in the Rhine Graben, with specific stress orientations, can be hydraulically conductive and therefore exploited for open EGS systems. Currently designed for small-scale local applications, these projects typically produce around 5 MWe, sufficient to heat around 10,000 homes. However, open EGS systems could be scaled up to ~100 MWe, depending on the temperature and flow rate found at great depth.

Although these small-scale projects have been successful, for the EGS concept to be universally applicable, one needs to test it away from the hydrological characteristics of the Rhine Graben.

Figure 9.3 Wheal Jane tin mine near Truro, Cornwall, UK. Courtesy of John Peck.

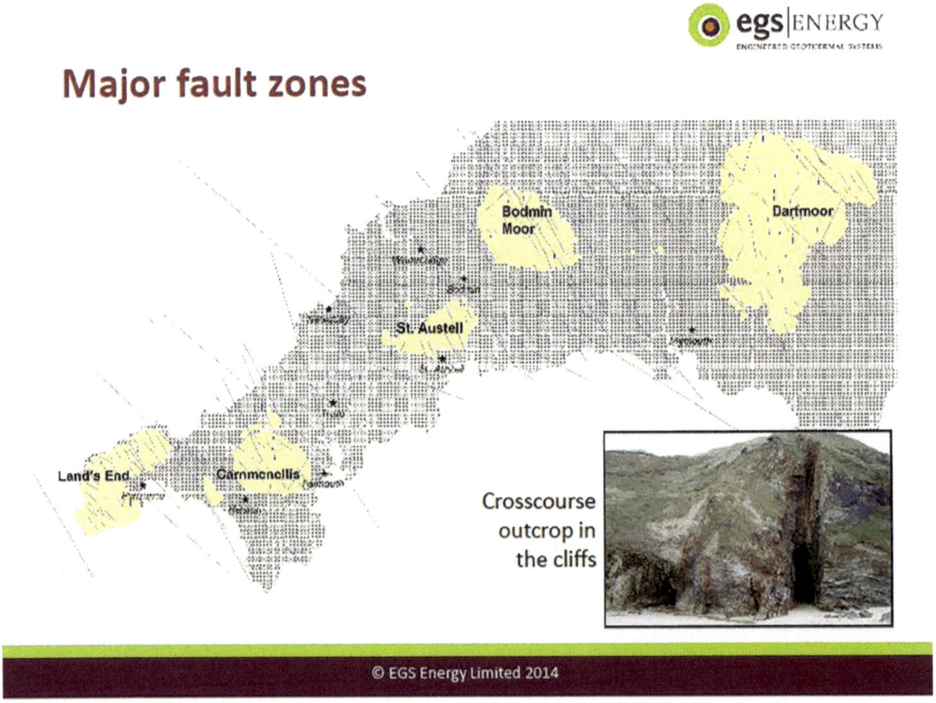

Figure 9.4 Satellite image and surface expression of faults (crosscourses) in SW England.

9.4 INTERNATIONAL COOPERATION ON EGS

Time is of the essence to mitigate the climate change issue. International cooperation and a systematic approach are proposed to make this approach a success. It is therefore proposed that a large-scale international collaborative project addressing EGS be undertaken, following the broad series of steps described below.

9.4.1 Identify EGS resources worldwide

It is proposed that international cooperation be established under a consortium consisting of the International Energy Agency, the World Bank, United Nations, individual state geological surveys and others to map/identify specific deep fault systems in all countries using their geologists where possible or seek external help/aid.

EGS resources closer to populated areas, where demand is higher should be given priority to demonstrate the value of EGS for mitigating climate change issues. However, several such projects should be undertaken simultaneously to ensure that the programme remains viable if there are unforeseen difficulties with any of the sites.

9.4.2 Reduce the risk of low-permeability wells

The next step would be to characterise deep fault systems using a combination of stress data (from the World Stress Map and other data) and geophysical techniques (including resistivity, gravity and seismic). The purpose would be to identify faults with the greatest inferred permeability. These faults could be characterised by their perceived properties, in order of decreasing risk:

a. Uncertain permeability
b. Moderate permeability
c. High permeability

This characterisation of deep faults would guide decisions about pursuing specific sites and the associated risks. In due course, and with significant international cooperation for data and knowledge sharing, experience will demonstrate the characteristics of successful EGS sites, to the benefit of everyone involved.

9.4.3 Improve heat utilisation

To maximise the value of energy produced from open EGS systems, the overall efficiency, performance and cost-effectiveness of above-ground systems need to be optimised, e.g., by

- extending the temperature range of different applications of geothermal energy;
- improving binary power plant efficiency; and
- improving the integration of geothermal heating, cooling and electricity supply into local energy systems.

The goal is to maximise the generation at the lowest life-time cost.

The layout of the plant and its usage will need to meet local requirements and may have to be tailored to a specific need of the local community. It would be an advantage to make information about the optimisation process available for international cooperation to build an inventory of best practises and thereby reduce risk, improve the return on investment and promote the uptake of this approach worldwide.

9.4.4 Improve drilling economics

Drilling in deep basements is one of the main stumbling blocks to the advancement of open EGS technology worldwide. Recent research and continuing development in percussion drilling suggest that significant improvements can be achieved using this technique when compared to traditional rotary drilling. Tests in Finland (https://www.st1.eu/st1-geothermal-pilot-project-otaniemi-espoo-reaches-stimulation-stage) have demonstrated that percussion drilling can improve the rates of penetration (ROP) by several fold compared to that of conventical rotary drilling, which can dramatically improve the economics of geothermal and make it acceptable worldwide. One of the uncertainties in percussion drilling is the depth to which it can work efficiently. The present limitation appears to be a depth of around 4,500 m, depending on geology, borehole diameters, etc. With continuing research and a greater uptake of percussion drilling by the geothermal industry, EGS technology may grow exponentially.

Development of a worldwide database on percussion drilling and other drilling methods suitable for EGS (together with information on geologic and stress conditions) would provide useful information to EGS developers for selecting the optimal drilling approach and reducing drilling costs as much as possible. Funding for improved deep geothermal drilling techniques (with the goal of increasing efficiency and reducing cost) is being provided in various countries; ideally, there would be coordination between countries to ensure that best practises are available and understood by the EGS development community.

One potential means of enhancing the economics of EGS resource development would be to include the co-production of lithium, cobalt and rare metals dissolved in the geothermal fluids, thus providing additional revenue. Although not directly related to drilling improvements, the addition of such a revenue stream could help justify drilling costs, which typically account for more than half of the costs of developing an EGS project.

9.4.5 Scaling up

A two- or three-well EGS system may produce anything from 5 to 15 MWe, depending on the per-well temperature and flow rate. This is satisfactory for providing small, local energy needs; however, for EGS to become a significant regional resource, scaling up to 100 MWe or more is required. This is not outside the bounds of current technology and could be attained using multi-well systems. As the understanding of the deep faults grows and the cost of geothermal drilling is lowered, scaling up the open EGS system will be the natural result, making geothermal a popular and friendly resource to be exploited worldwide for commercial-scale operation by both private- and public-sector developers.

9.4.6 Establish a legal and regulatory framework and increase social acceptance

Introducing and deploying deep geothermal technologies on a large scale has a number of technical and non-technical challenges, notably:

- high investment costs
- the acceptance of the technologies in the developing areas and
- the involvement of consumers in the production chain to enhance the role of energy distribution.

In parallel with the technological research described above, coordinated actions to develop regulatory, financial, political and social solutions would help overcome such barriers and accelerate the deployment of EGS on a global scale, enabling it to become one of the main mitigators of global climate change and meet energy demand targets.

9.5 DEVELOPMENT OF AN EGS ACTION PLAN

The following sections outline a step-by-step approach for an EGS Action Plan.

9.5.1 EGS Centre of Excellence

Previous experience at the European EGS project at Soultz, France, has shown that interaction and cross-fertilisation of ideas between scientists and engineers within the international community helped significantly in the development of the technology. Significant progress in the technology and interpretation of the data was achieved within the European EGS project, which reduced the risk and improved the technology.

As described above, Cornwall is renowned for the richness of its natural mineral resources. The county's history is embedded in the mining of valuable economic commodities dating back to Roman times. This tradition would enable a new wave of activity: using EGS technology to recover a vast resource of stored heat in the granite of the Cornubian Batholith. Estimates (SKM and REA, 2012) indicate that by accessing hot rocks at 5,000 m, up to 20% of the current electricity generation capacity of the UK can be delivered for up to 200.

Using a similar concept as at Soultz, a project can be initiated 1,200 km away at a possible site in the southern part of Cornwall. As described above, geologic and stress conditions are favourable and there is a significant body of knowledge about conductive faults and fractures. Temperature modelling (Willis-Richard et al., 1990) indicates a temperature of around 220°C at 5 km depth. A successful EGS project at this site would demonstrate that these deep fault systems resemble hydrothermal systems, but in deep igneous rock away from Rhine Graben and at other similar locations identified as described in Section 9.4.1, which will demonstrate the accessibility of open EGS systems in many parts of the world.

It is proposed that an EGS project in Cornwall would become the basis for the initial international Centre of Excellence for EGS technology. The centre would be operated in conjunction with private- and public-sector organisations, including geothermal developers, mining companies, oil and gas companies, two exceptional local academic institutions (Universities of Exeter and Bristol) and academic institutions from other countries. A high level of participation will be encouraged in this venture to ensure that technology can be transferred globally. It is anticipated that the Centre of Excellence will have a close link with the IEA, World Bank and UNESCO, ensuring the effective application of open EGS systems to address the mitigation of climate change.

Another engineering topic that could be pursued at this centre would be the use of geothermal heat for industrial and domestic applications. It is well understood that a significant CO_2 reduction can be achieved if one can replace the industrial use of heat from hydrocarbon energy with that from geothermal energy, thus moving towards climate mitigation support. It is expected that local academic institutions will develop graduate and postgraduate courses in geothermal engineering to facilitate the worldwide deployment of open EGS technology.

Once the concept and structure of the first Centre of Excellence are established, they can be replicated in many parts of the world to facilitate access to regional specificity and address any specific geological conditions and related topics.

9.5.2 Global funding and adoption of EGS

The initial financing of demonstration EGS projects in many countries may be too large a financial commitment in some locations. A possible mechanism to address this issue would be to involve public financial organisations such as the World Bank. The relatively small investment in advancing EGS deployment should be compared with the dire economic predictions if nothing is done to mitigate climate change. Once the concept has been demonstrated in a specific country, a mechanism could be put in place for it to be taken up by state or private entities for further development. The authors believe that this is an international issue, and continued international cooperation is the only way forward. This may require setting up a specific organisation for the advancement of EGS energy, perhaps under established organisations such as the United Nations, the World Bank, or other international institutions, with the goal of worldwide deployment to demonstrate that this technology is for all countries, regardless of their development status.

Each Centre of Excellence would incorporate a long-term funding strategy for R&D to continually improve technology, reduce risk and enhance economics, as well as a public outreach campaign to enhance public perception of EGS and thus acceptance of it worldwide.

By demonstrating the economics of EGS schemes, there will be a commercial case for investment from companies currently focused on fossil fuels, as their operational strategies already embrace the thinking behind resource optimisation. In addition, EGS would provide an added benefit to these stakeholders: unlike a barrel of oil or a ton of coal, they can sell the energy they produce continuously over a long period.

The successful adoption of EGS by an appreciable number of nations would make mitigation efforts for climate change more successful and the future for humankind more positive.

9.5.3 Global cooperation, data collation and data sharing

Regular international meetings where various parties are brought together to discuss the current state of specific projects would be highly beneficial for realising a future with EGS. Such meetings could be held in addition to existing, well-attended geothermal meetings, such as the annual Stanford Geothermal Workshop and other meetings such as the World Geothermal Congress, the Geothermal Resources Council Annual Meeting and conferences held by the European Geothermal Energy Council (EGEC). A specific website and other information-sharing methods would be developed to enable cross-fertilisation between projects by hosting discussion forums and sharing common problems and solutions. Sharing of human resources, knowledge and materials should be encouraged.

9.6 CONCLUSIONS

Open EGS resources represent not only an unprecedented opportunity/mechanism to mitigate climate change issues with renewable energy sources but would also help developed and developing nations to work together to access an indigenous and ubiquitous energy resource, improving the quality of life, reducing tensions between nations and sharing the prosperity that EGS can bring.

Overall, the global geothermal potential (hydrothermal + EGS) is huge and renewable. It is an environmentally friendly, strategic global resource well suited to supply baseload power and offset the demand for electricity. By following a path similar to that outlined herein, EGS can become economically attractive and a predominant energy resource for the world in the future.

ACKNOWLEDGEMENT

The authors would like to thank international funding agencies and colleagues who have contributed so much of their professional lives and time to provide an improved understanding of EGS systems. We are especially grateful to Ladsi Rybach, Fritz Rummel, Francois Cornet, Don Brown, Subir Sanyal, Roland Horne, Philippe Dumas, Lothar Wissing, John Garnish, Allen Jelacic, Terry Gandy, Hiroki Niitsuma and others who have an exemplary record of their dedication to EGS.

The authors would like to give special acknowledgement to Perry Moore, a genius for drilling in hard rocks, who passed away on 21 March 2003 in Kazakhstan. He had absolute faith in the EGS concept from his Los Alamos days and supported the programme with tremendous drive and enthusiasm.

REFERENCES

Abe, H., Niitsuma, H., Baria, R. (1999). Hot dry rock/hot wet rock academic review: special issue. *Geothermics* 28 (4/5), 0410, ISSN 0375–6505.

Baria, R., (ed.). (1990). Hot dry rock geothermal energy, *Proceedings of Camborne School of Mines, International HDR Conference, June 1989*, Robertson Scientific Publications, London, 613 pp.

Baria, R., Garnish, J., Baumgaertner, J., Gerard, A., Jung, R. (1995). Recent development in the European HDR research programme at Soultz-Sous-Foret (France), *Proceedings of the World Geothermal Congress, Florence, Italy*, International Geothermal Association, Vol. 4, pp. 2631–2637, ISBN 0-473-03123-X.

Baria, R., Green, A.S.P. (1990). Microseismic: A key to understanding reservoir growth, in R. Baria (Ed.), *Hot Dry Rock Geothermal Energy, Proceedings of the Camborne School of Mines International Hot Dry Rock Conference, Camborne School of Mines, Redruth, UK, 27–30 June, 1989*, Robertson Scientific Publications, London, pp. 363–377.

Baria, R., Baumgärtner, J., Rummel, F., Pine, R.J., Sato, Y. (1999). HDR/HWR reservoirs: concepts, understanding and creation. *Geothermics* 28 (4/5), 533–552, ISSN 0-375-6505.

Baria, R., Baumgaertner, J., Gerard, A., Garnish, J. (2000). The European HDR programme: Main targets and results of the deepening of the well GPK2 to 5000 m, *World Geothermal Congress 2000, Kyushu - Tohoku, Japan.*

Baria, R., Baumgaertner, J., Glass, H., Bennett, T., Teza, T., Jupe, A. (2016). Unconventional geothermal technology to help the climate change issue in the densely populated areas of the world with a demand for higher energy, *European Geothermal Congress, 19–24 September 2016, Strasbourg, France.*

Batchelor, A.S. (1990). Hot Dry Rock and its relationship to existing geothermal systems, in R. Baria (Ed.), *Proceedings of the Camborne School of Mines International Hot Dry Rock Conference*, Camborne School of Mines, Redruth, UK, 27±30 June. Robertson Scienti®c Publications, London, pp. 13±29.

Baumgärtner, J., Moore, P., Gérard, A. (1995). Drilling of hot and fractured granite at Soultz, *Proceeding of the World Geothermal Congress, Florence, Italy.* International Geothermal Association, Vol. 4, pp. 2657–26663, ISBN 0-47303123-X.

Baumgärtner, J., Gérard, A., Baria, R., Jung, R., Tran-viet, T., Gandy, T., Aquilina, L., Garnish, J. (1998). Circulating the HDR reservoir at Soultz: maintaining production and injection flow in complete balance, in *Proceedings of the 23rd Workshop on Geoth. Reservoir Engineering.* Stanford University, Stanford, CA.

Baumgärtner, J., Lerch, C. (2013). Geothermal 2.0: The Insheim geothermal power plant. The second generation of geothermal power plants in the Upper Rhine Graben, *Third European Geothermal Review, Geothermal Energy for Power Production, June 24–26, 2013, Mainz, Germany.*

Baumgärtner, J., Teza, D., Wahl, G. (2013a). Gewinnung geothermischer Energie durch Entwicklung und Zirkulation eines Störungssystems im Kristallin und deren mikroseismische Überwachung am Beispiel des Geothermieprojektes Insheim (Extraction of geothermal energy through the development and circulation of controlled systems in the crystalline basement and their microseismic monitoring, using the example of the geothermal project Insheim). Final report to BESTEC GmbH. Accessed online 4 February 2016. http://edok01.tib.uni-hannover.de/edoks/e01fb14/777426781.pdf.

Baumgärtner, J., Hettkamp, T., Teza, D., Kölbel, T., Mergner, H., Schlagermann, P., Lerch, C. (2013b). Betriebserfahrungen mit den Geothermiekraftwerken Landau, Insheim und Bruchsal. bbr Fachmagazin für Brunnen- und Leitungsbau.

Bertani, R. (2010). Geothermal power generation in the world 2005–2010. Update report, *Proceedings at World Geothermal Congress 2010, 25–29 April 2010, Bali, Indonesia.*

Evans, K.F. (2005). Permeability creation and damage due to massive fluid injections into granite at 3.5 km at Soultz: part 2 – critical stress and fracture strength. *Journal Geophysical Research* 110, 14. doi:10.1029/2004JB003169.

Garnett, R.H.T. (1961). Structural control of mineralization in South-West England. *Mining Magazine* 105, 329–337.

Garnish, J.D. (1976). *Geothermal Energy: The Case for Research in the United Kingdom.* Energy paper no. 9, HMSO, London, 66 pp.

Gérard, A., Baumgärtner, J., Baria, R. (1997). An attempt towards a conceptual model derived from 1993–1996 hydraulic operations at Soultz, in *Proceedings of NEDO International Geothermal Symposium, Sendai*, Vol. 2, pp. 329–341.

Hawkes, J.R. (1981). A tectonic 'watershed' of fundamental consequence in the post-Westphalian evolution of Cornubia. *Proceedings of the Ussher Society* 5, 128–131.

Hosking, K.F.G. (1949). Fissure systems and mineralisation in Cornwall. *Transactions of the Royal Geological Society of Cornwall* 18, 9–49.

International Energy Agency. (2011). *IEA Road Map: Geothermal Heat & Power.* Accessed at http://iea-gia.org/wp-content/uploads/2012/08/Geothermal_Roadmap-Final-Pub-Ver-15Jun11.pdf.

International Energy Agency. (2009). *IEA World Energy Outlook 2009, Climate Change Excerpt.* International Energy Agency, Paris.

Krewitt, W., Nienhaus, K., Klessmann, C., Capone, C., Stricker, E., Graus, W., Hoogwijk, M., Supersberger, N., von Winterfeld, U., Samadi, S. (2009). *Role and Potential of Renewable Energy and Energy Efficiency for Global Energy Supply.* Federal Environment Agency (Umweltbundesamt), Dessau-Rosslau, Germany. Available at www.umweltbundesamt.de/uba-infomedien/mysql_medien.php?anfrage=Kennummer& Suchwort=3768.

Moriya, H., Niitsuma, H., Baria, R. (2003). Multiplet-clustering analysis reveals structural details within the seismic cloud at the Soultz geothermal field, France. *Bulletin of the Seismological Society of America* 93 (4), 1606–1620.

Mueller, S., Scott, P.W., Evans, M.J. (1999). Kaolinisation, mineralisation and structures in biotite granite at Bodelva, St Austell, Cornwall. *Geoscience in South West England* 9, 310–317.

Parker, R.H. (1989). *Hot Dry Rock Geothermal Energy: Phase 2B Final Report of the Camborne School of Mines Project.* Pergamon Press, Oxford, UK, 1391 pp.

Pine, R.J., Baria, R., et al. (1987). A technical summary of phase 2B of the Camborne School of Mines HDR project. *Geothermics* 16 (4), 341–353.

Scrivener, R.C., Darbshire, D.P.F., Shepherd, T.J. (1994). Timing and significance of crosscourse mineralization in SW England. *Journal of the Geological Society of London* 150, 587–590.

Shail, R.K., Alexander, A.C. (1997). Late Carboniferous to Triassic reactivation of Variscan basement in the western English Channel: Evidence from onshore exposures in south Cornwall. *Journal of the Geological Society* 154, 163–168.

SKM (Sinclair Knight Merz), REA (Renewable Energy Association). (2012). *Geothermal Energy Potential, Great Britain and Northern Ireland.* Accessed on the REA website at: https://www.r-e-a.net/upload/skm_report_on_the_potential_for_geothermal_energy_in_gb__ni_may_2012.pdf.

Soma, N., Niitsuma, H., Baria, R. (2002). Evaluation of subsurface structure at Soultz hot dry rock site by the AE reflection method in time-frequency domain. *Pure Applied Geophysics* 159, 543–562.

Valley, B. (2007). The relation between natural fracturing and stress heterogeneities in deep-seated crystalline rocks at Soultz-sous-Forêts (France). Doctoral Thesis. Swiss Federal Institute of Technology (ETH), Zurich, Switzerland, 241 pp.

Willis-Richard, J., Thomas-Betts, A., Sams, M., Wheildon, J. (1990). HDR resource in South West England, in R. Baria (Ed.), *Hot dry rock geothermal energy, Proc. Camborne School of Mines International Hot Dry Rock Conference, Camborne School of Mines Redruth, UK, 27–30 June, 1989,* Robertson Scientific Publications, London, pp. 215–223.

Index